MAPPING REALITY

Mapping Reality

An Exploration of Cultural Cartographies

Geoff King

St. Martin's Press
New York

MAPPING REALITY

For information, address:

St. Martin's Press, Scholarly and Reference Division,
175 Fifth Avenue, New York, N.Y. 10010

First published in the United States of America in 1996

Printed in Great Britain

ISBN 0–312–12704–9 (cloth)
ISBN 0–312–12706–5 (paper)

Library of Congress Cataloging-in-Publication Data applied for

For Alison, with thanks to Geoff Hemstedt

Contents

1 The Map that Precedes the Territory

Henceforth, it is the map that precedes the territory [. . .].
Jean Baudrillard[1]

The midwestern community of Lake Wobegone in Garrison Keillor's novel *Lake Wobegone Days* (1987) does not exist on the map. Mistakes were made by early cartographers working without the benefit of aerial views or modern technology. The map they produced did not quite fit the territory and a section of central Minnesota had to be erased. The map was meant to be provisional but its lines came to be inked in and Lake Wobegone was lost along with the rest of the appropriately named Mist County. Some citizens attempt to overcome the problem of their cartographic non-existence. A legislative commission proposes that the missing land be reinstated by reducing the space occupied on the map by a number of lakes: the lakes could be elongated to ensure that no valuable shoreline property is lost. Opposition to the proposal is led by the Bureau of Fisheries, which is concerned that the reduction in the size of the lakes would mean the loss of valuable walleye breeding grounds, and the proposal is lost. The whole community is split over the issue:

> Proponents of map change, or 'accurates' as they were called, were chastised by their opponents, the so-called 'moderates', who denied the existence of Mist County on the one hand – 'where is it?' a moderate cried one day on the Senate floor in St. Paul. 'Can you show me one scintilla of evidence that it exists?' – and, on the other hand, denounced the county as a threat to property owners everywhere. 'If this county is allowed to rear its head, then no boundary is sacred, no deed is certain,' the moderates said. 'We might as well reopen negotiations with the Indians.'[2]

The map, as Jean Baudrillard puts it, has come to precede the territory. Rather than the map being a product *of* the territory, as it is usually understood, coming only after it – both temporally and conceptually – and remaining answerable to it, there has been a curious reversal. The debate is conducted in terms of the map rather than the territory itself. Of course, the walleye breeding grounds would still exist physically and the actual lakes would be undiminished in size if the map alone were altered, just as Lake Wobegone continues to exist off the map. But the map is more than merely a passive representation of the territory. That which is marked on the map is affirmed as real and changes to the map are important matters. The fishermen are unlikely to be affected by the removal of their grounds from the map, we might think. The fish would still breed and still be caught. So why the objection? Might the map have so great an authority, an apparent objectivity, as to convince even the locals or to question their own judgement or experience? Others might not in future appreciate the reserves – potential markets, perhaps, or whoever monitors such matters on a state or federal basis. Or is it just the principle? As Keillor's moderate suggests, one change to the map would open up the possibility of others. If the people of Lake Wobegone can have the map redrawn to suit their own needs others might apply similar pressure, with who knows what consequences. The other lines on the map might prove equally arbitrary.

But what about the fate of a place like Lake Wobegone, absent from the map? The consequences could be serious. Who is likely to move to, invest in or even visit a place that is denied cartographic representation? Similar concerns were voiced recently in the neighbouring state of North Dakota. Newspaper columnists reacted with outrage to a Rand McNally atlas on which their territory received only diminutive representation. They were being written off the map, they complained, in what seemed to be a further manifestation of the social and economic decline of the region. The people of North Dakota might have something in common with those of Peter Carey's story 'Do You Love Me?' (1981), whose concern with the mapping of their territory is grave. Cartographers are the most important members of a community that craves to know the exact shape and extent of its uncertain land. Mapping provides an affirmation of their existence. When parts of the world begin to dematerialize the role of the cartographers becomes acute. A device known as the Fischerscope is developed to detect even those parts that have been lost:

> In this way the Cartographers were still able to map the questionable parts of the nether regions. To have returned with blanks on the maps would have created such public anxiety that no one dared think what it might do to the stability of our society. I now have reason to believe that certain areas of the country disappeared so completely that even the Fischerscope could not detect them and the Cartographers, acting under political pressure, used old maps to fake-in the missing sections.[3]

Here the map has become the only reality, not representing a territory but establishing its sole form of existence. It is tempting to suggest something of a parallel in the expressions of outrage that greeted plans in November 1988 for the sale by Hereford cathedral of its precious *mappa mundi*, a rare thirteenth-century depiction of the world. Questions were asked in Parliament, petitions were signed and the map was propelled from obscurity into the national limelight before eventually being rescued with last-minute donations from National Heritage and John Paul Getty Jr. A cynic might have wondered if as much concern was being expressed for the fate of the territory itself at a time of heightened global environmental destruction.

A similar sense of a confused or inverted hierarchy of relations between map and territory is given by Jeannette Winterson in *Sexing the Cherry* (1989): 'A map can tell me how to find a place I have not seen but have often imagined,' she suggests. But: 'When I get there, following the map faithfully, the place is not the place of my imagination. Maps, growing ever more real, are much less true.'[4] This might be the case for the modern reader of R.D. Blackmore's romantic saga, *Lorna Doone* (1869). The mythical Doone valley turned up on a recent Ordnance Survey map of Exmoor. The Ordnance Survey thought it was doing the reading public a favour, pinning down the spot in which the romance unfolded. Not so for the American Lorna Doone Society, which complained that it had all happened in a different valley: 'They want the map reprinted; a rival group in the UK does not. And Ordnance Survey men who were despatched to Exmoor, armed with annotated copies of Lorna Doone as well as their theodolytes and tape measures, have been unable to settle the argument.'[5] Many other landscapes have also been read according to literary mappings that can become confused with reality, including those of 'Hardy's Wessex', 'Brontë country', and 'Herriot's Yorkshire', as the tourist industry has been quick to ensure on its maps.

The fictional, once mapped, may become real. Or the territory may be altered to fit the map, as in the Walter Abish novel *Alphabetical Africa* (1974) where the population of an imaginary Tanzania is kept in employment in the Sisyphean task of painting the entire country orange to conform with its colour on the map. Alternatively, the map may physically outgrow the territory. A character in Lewis Carroll's *Sylvie and Bruno Concluded* (1894) tells of a map made on a scale of a mile to the mile. It had never been used because of objections from farmers: 'They said it would cover the whole country, and shut out the sunlight! So we now use the country itself, as its own map, and I assure you it does nearly as well.'[6] Nearly, but not quite. The territory does not represent itself quite as well as does the map. Yet the map would obliterate or even replace the territory. This is precisely what happens in the Jorge Luis Borges fragment from which Baudrillard draws the image of the map that precedes the territory. Cartography reaches a state of such perfection that a map is produced on the same scale as the empire it represents. Rejected by later generations as too cumbersome 'and not without Irreverence',[7] it is abandoned in the desert, only a few tattered fragments surviving the rigours of sun and rain. For Baudrillard the outcome is to be reversed:

> It is the map that engenders the territory and if we were to revive the fable today, it would be the territory whose shreds are slowly rotting across the map. It is the real, and not the map, whose vestiges subsist here and there, in the deserts which are no longer those of the Empire, but our own. *The desert of the real itself.*[8]

This is seen as an allegory of a state of cultural simulation in which traditional distinctions between map and territory, image and reality, are no longer tenable:

> It is no longer a question of either maps or territory. Something has disappeared: the sovereign difference between them that was the abstraction's charm. For it is the difference which forms the poetry of the map and the charm of

> the territory, the magic of the concept and the charm of the real. This representational imaginary, which both culminates in and is engulfed by the cartographer's mad project of an ideal coextensivity between the map and the territory, disappears with simulation [. . .].[9]

In later works Baudrillard comes explicitly to associate this blurring of distinctions between map and territory with the experience of the 'postmodern', as have a number of others. According to Scott Lash, for example, the era of modernism was characterized by an increasing process of differentiation in which various spheres of social and cultural life became increasingly autonomous. The postmodern, then, is a process of *de*-differentiation, akin to Baudrillard or Marshall McLuhan's notion of a carnivalesque 'implosion' of previously differentiated categories.[10] This has been described in various ways. In the wake of assorted developments in electronic and other media, it is often argued, the reality we inhabit in Western society is increasingly comprised of images, to a point at which image and reality can no longer adequately be distinguished.

From a traditional realist perspective neither representation nor the reality represented was viewed as problematic. A solid reality existed of which adequate representations could easily be produced; the territory could be mapped without problem. This is of course an over-simplified account. Definitions of the real have always been subject to contest. In the modernist experience, it is argued, the notion of representation in general came under more concerted question. As representational forms became more autonomous they began to take on a distinct opacity and became disconnected from everyday reality. In the postmodern it is reality itself that is said to have become problematic. Representations, formerly understood as secondary elements, become central to the fabric of our lives.

For Baudrillard, this process has reached a stage in which any notion of the real as something existing in its own right has become untenable: 'The very definition of the real becomes: *that of which it is possible to give an equivalent reproduction.*' Further: 'the real is not only what can be reproduced, but *that which is always already reproduced.* The hyperreal.'[11] Reality itself is something that appears to be both more and less real in a world in which inversions effected on a local basis by the surrealists and limited to the world of the imagination have become general: 'Today it is the quotidian reality in its entirety – political, social, historical, and economic – that from now on incorporates the simulatory dimension of hyperrealism.'[12] More than anywhere else, we are told, this hyperreal dimension is to be found in the world of television, that familiar presence in millions of homes that both reflects and shapes the spaces in front of the screen. The psychoanalyst Jacques Lacan's 'mirror phase', in which our sense of individual identity is said to be gained through the reflections provided by others, 'has given way to the video phase'.[13] Interactions with real others are being replaced by pervading electronic mediations of basic experience, whether in broadcast television, videotape, the amusement arcade, the Nintendo game, electronic personal organizers or other computers used in the home, business or school. The computer Internet opens up a new landscape of international 'cyberspace' the uses and implications of which are only beginning to be explored.

The television viewer is said to be more 'inside' the world behind the screen than the filmgoer, for whom the distinction between fiction and reality is more explicit.[14] Reality becomes 'just another channel', an idea that finds its ultimate expression in the world of computer-generated virtual realities that offer the prospect of something close to a total immersion in the medium. One of the early products was a virtual map of the streets of Aspen, Colorado, created at the Massachusetts Institute of Technology. Footage taken by a mobile camera system enabled the user to experience the illusion of moving around at will inside the reality of the map.[15] Virtual reality technologies are at an early stage. The worlds they conjure are crude and often cartoon-like, although the advent of new generations of computers is likely to create the possibility of a new dimension of entertainment, a development already foreshadowed at Disneyland and elsewhere. Some fantasize the possibility of a world in which actual could no longer adequately be distinguished from virtual reality, or where virtual worlds might prove so seductive, even addictive, that we might never want to return from such ordered or ideal environments to the complications of the world outside. It might be argued that this is already the case for some of those seduced into the worlds of their favourite forms of electronic entertainment.

This is precisely how Baudrillard envisages a postmodern landscape in which there is no reality underlying the virtual map. It was a typical Baudrillard intervention to dismiss the war in the Arabian Gulf in 1991 as a form of virtual reality, an 'unreal war, war without the symptoms of war, a form of war which means never needing to face up to war, which enables war to be perceived from deep within a darkroom.'[16] Before, during and even after the war, he continued to doubt its real-world credentials.

The war did happen of course, and could hardly have been more real for those in Iraq on the receiving end of one of the heaviest bombardments in history. It also had very real implications for the future political map of the Middle East. Virtual and televisual images may have played an unusually large part in the conflict; the point that should be stressed is that such images were central players *in* the reality of the war rather than in some way effacing it. Where bomber crews saw not the actual target but its image on screen, or so-called smart weapons followed computer-generated maps to reach their destinations, the destruction created on the ground was hardly any less real. Screenings of selected images of laser-guided bombs hitting their targets with pinpoint accuracy were important ideological tools in the maintenance of support for the assault on Iraq, not evidence that the war existed only in the realm of television spectacle. They gave the impression that what was involved was a precise 'surgical' strike, although most of the weapons used were more conventional, less accurate and led to civilian death and destruction. Images of American bombs hitting their targets with such apparent accuracy also served a useful purpose of deterrent, a warning for the future if not for the immediate conflict of a capacity for precise destruction that might put the lie to memories of Vietnam, where high technology weaponry failed to achieve the kind of success that had been expected, or more recent assaults on Libya and Panama in which 'smart' weapons had failed many of their examinations.

Control over the production of images of the conflict was an important strategic issue for an American administration seeking to hold together a variety of fragile supporting coalitions. The advent of new video technologies both widened the audience for official images and narrowed the range of perspectives on offer to the general viewer. The hegemony of television images has increased with the development of video-still equipment used to transplant the video image onto the pages of newspapers. Individual frames can be 'grabbed' electronically and the image reproduced without the need for time-consuming chemical processing. Newspapers can quickly gain images of events such as the Gulf war that are often accessible only through officially released televisual coverage. Borrowing from the apparent live presence of the television broadcast, such images may at times have a greater rhetorical power than conventional newspaper photographs: 'This "video effect" (in which images are broken down into horizontal lines, sometimes with a black border) has become an index of "immediacy", replacing the 35 mm blurred black-and-white picture as signifying the dramatic news event.'[17] In a world in which many events seem to possess their greatest reality on the screen, it is logical that the photograph of the screen, the image of the image, might become the signifier of the objectively real. The ultimate in reality might be suggested not by the presence of the newspaper photographer at the event but the appearance of the CNN logo in the corner of the picture. Such an identification of the picture as the property of a rival news service would traditionally be avoided but might in this case be valued by the newspaper frame-grabber as a talisman of immediacy.

Where both broadcasters and publishers become dependent on officially released images of this kind the subsequent reduction in the number of available points of view can have potentially totalitarian implications. The Gulf war is a case in point. Limited perspectives may also have been given of earlier conflicts such as the American intervention in Vietnam, but the coverage of that war was positively anarchic compared with the rigid controls imposed on those attempting to report from the Gulf. In Vietnam reporters were given relative freedom to move around the country as they chose; in the Gulf they were subject to tight restrictions and likely to be deprived of their accreditation if guilty of any breach of guidelines. Almost without exception the images that were reproduced were those given official sanction for release. Of the unknown thousands who died in Iraq there was little or no representation. When some such images did appear, particularly after the destruction of a bomb shelter in Baghdad, they had an effect on public opinion that underlined the need of the military to keep the broadcast of such scenes to a minimum.[18] Images were used to displace the impact of the reality of the war, rather than themselves effacing the reality of what happened. The distinction is an important one.

Video-still images can be manipulated in almost any manner without leaving any final trace of the process. Once the image has been captured and produced on a computer screen any number of alterations can be made. Elements can be added or removed; colour, focus, and exposure-effects can all be changed after the fact.[19] Such alterations leave no photographic trace. There is no original negative to which appeal can be made either to question or validate the authenticity of the image. The manipulator of the video-still image can in effect choose the reality that is to be represented. Video-still photography has thus been greeted by some as signalling the end of the role of the photograph in providing evidence of truth or reality. But photographic techniques have never offered anything more than their own perspectives on events, perspectives that might be all the more illusory the more they are believed to be objective. The standard 50 mm lens (with 35 mm stock) reproduces a view of the world similar to that constructed through the lens of the human eye. It appears to validate as real a prior reproduction of the world. A representation is reproduced rather than reality. The same might be said of the whole world of cultural cartography, although in the process the distinction between reality and representation becomes blurred.

Video-still images can be flashed instantly across the globe, helping to increase a blurring of boundaries between different cultures and places that has also been associated with the postmodern experience. According to one view this is the direct outcome of the development of global telecommunications and computer networks. Others associate the process more closely with economic factors such as the dominant position of multinational corporations that transcend the nations in which they are based. Financial, commodity and labour markets have become increasingly global. Money thus becomes 'an electronic impulse fibrillating across the world',[20] unhindered by distinctions of time, place, currency or commodity. The real, productive economy seems to have been eclipsed by parasitic wheeler-dealers at their computer terminals in New York, Tokyo and London. The world of financial and commodity markets may appear to be disconnected from the physical production and exchange of goods, but this again is not the move away from reality that the likes of Baudrillard suggest. It impinges heavily upon such realities. Whole regions may find their staple exports devalued overnight according to the dictates of the market.

Los Angeles is often cited as the embodiment of postmodern confusions between map and territory, both as home of much of the global entertainment industry and in terms of is own strange geometry. America more generally for Baudrillard is the acme of the postmodern, 'a tactile, fragile, mobile, superficial culture', that dominates the world on the basis of its fictions.[21] Los Angeles is a microcosm of the world, according to Edward Soja's *Postmodern Geographies* (1989). From the futuristic technologies of the military-industrial complex to the waves of exploited immigrant labour that allow parts of the city to compete with new industrial concentrations in the 'Third World', the global is localized and the local globalized.[22] The city is a decentred collection of suburbs notorious for its lack of focus, mapped only by a grid of freeways. The city, we are told, is 'hard to envision or conceptualise as a whole.'[23] The names of some of its parts – Venice, Naples, Westminster and others – attempt to recapture the kind of romanticized history and well-defined structure it lacks, 'making the lived experience of the urban increasingly vicarious, screened through *simulacra*, those exact copies for which the real originals have been lost.'[24] This is the kind of postmodern territory that Umberto Eco seeks to map in his essay 'Travels in Hyperreality' (1973). In the simulated landscapes around Los Angeles, Eco finds an assortment of museums and theme parks in which real and imaginary worlds begin to merge into one, a phenomenon that in American culture seems to spread beyond the walls of such enclosures and into the whole culture of consumerism.

But there is more to Los Angeles than the fluid, unmappable surfaces described (or celebrated) by theorists of the postmodern. It is also a city of rigid cultural mappings and demarcations, none more overt than those between the fortress-like white suburbs and the impoverished black ghettos 'written off' both the tourist map and the dominant image of the city broadcast by its own dream-machine. For many the South Central district that forced its way onto the map during the riots of April 1992 had been either repressed from consciousness or viewed as a dangerous wilderness to be crossed only on the elevated freeways that guard against any contact with its threatening reality. The riots (following the acquittal of police officers accused of beating Rodney King, a black motorist) came just as the city was preparing itself for an expected earthquake, a vivid reminder that the ground on which it was built was cultural rather than natural, the subterranean gulfs a matter of race and class as much as geological plate tectonics. If the city stands as a metaphor for a world of confusing facades and illusions, as it is presented in Nathanael West's *The Day of the Locust* (1939), it also suggests the violence underpinning such constructions; usually suppressed but provoking periodic outbursts such as the 1992 riots or the apocalyptic climax of the novel in which reality blurs into the painted image of 'The Burning of Los Angeles'. During the riots a new map of the city was flashed across the world. Familiar markers locating the homes of the stars were replaced by flame symbols charting the spread of an uprising that put the lie to any suggestion that the prevailing American reality was one of equality and freedom.

The dominant mapping of Los Angeles is a repressive grid that inscribes lines of class and race oppression onto the territory. New neighbourhood orientations were deliberately mapped onto the changed social landscape of the immediate postwar period in an effort to avoid disorientation.[25] Many other boundaries have since been imposed by wealthy white communities seeking to maintain their exclusivity. A range of zoning regulations and other devices have been used to protect privileged enclaves from unwelcome intrusions. Directly implicated in this process, Mike Davis suggests, are the structures at the heart of 'postmodern' Los Angeles. The Bonaventure Hotel, a key landmark in the cartography of postmodernism, is part of a fortress downtown redevelopment designed to exclude the homeless, the unemployed or the working poor in an assertion of security that followed the Watts rebellion of 1965. Luxury developments outside the city have been converted into walled enclaves with their own security services and privatized roads that are off-limit to those without residential passports. For Fredric Jameson, the topography of the Bonaventure is typical of a disorienting postmodern space that transcends 'the capacity of the individual human body to locate itself, to organize its immediate surroundings perceptually, and cognitively to map itself in a mappable external world.'[26] But the hotel and its pristine shopping mall complex is a structure that allows precisely for a mapping conducted according to strict lines of class and race. It is a territory mapped out for the occupation of the white middle and upper classes, sealed off from the nearby poor neighbourhoods. The Bonaventure may not provide a suitable metaphor for the elusive spaces of 'postmodern' communications and computer technologies, tools that have themselves been used to help police the mapping of the city's interior borders.[27]

Jameson's emphasis is less on the technology itself than its role in the world system of multinational or 'late' capitalism, as defined by Ernest Mandel.[28] The concept of cognitive mapping is drawn from Kevin Lynch's account of cityscapes such as Los Angeles and Jersey City – with their alienating structures, supposedly difficult to map other than in small parts – and more comprehensible or 'imageable' urban territories such as Boston. Lynch describes only a limited form of mapping, however: the traveller's subject-centred diagram of his or her own journey. In the history of mapping, as Jameson suggests, this might be equated with the limitations of medieval portolan charts used to map relatively small areas to aid navigation at sea. A satisfactory cognitive mapping would be more ambitious. It might be likened to the cartographical practices that followed the development of surveying instruments such as the compass, sextant and theodolite. Here we find 'a whole new coordinate – that of the relationship to the totality, particularly as it is mediated by the stars and by new operations like that of triangulation.'[29] New and abstract factors are brought into play alongside the empirical position of the subject. A further level of cognitive mapping might be the equivalent of the construction of the first globes and the invention of the Mercator projection of the world. This would add a third dimension of cartography:

> which at once involves what we would today call the nature of representational codes, the intricate structures of the various media, the intervention, into naive mimetic conceptions

> of mapping, of the whole new fundamental question of the languages of representation itself; and in particular the unresolvable (well-nigh Heisenbergian) dilemma of the transfer of curved space to flat charts; at which point it becomes clear that there can be no true maps (at the same time it also becomes clear that there can be scientific progress, or better still, a dialectical advance, in the various historical moments of map-making).[30]

The question of improvements in mapping techniques and the extent to which they represent increasing objectivity will be considered in the next chapter. Whatever improvements may have been made, Jameson suggests, no effective cognitive mapping of the postmodern experience has been achieved. The difficulty, he argues, is one that has arisen from changes in the structure of capitalism. In a reductive and over-simplified mapping of his own, Jameson neatly associates stages of realist, modernist and postmodernist representation with, respectively, market, monopoly and late capitalism. In the stage of market capitalism, he suggests, the immediate and limited experience of individuals was still able to encompass and coincide with the underlying social and economic forms that governed them. Under monopoly capitalism, or the stage of imperialism described by Lenin, the two levels are said to have drifted further apart, opening up a rift between essence and appearance, structure and lived experience. The structural factors of an increasingly international and imperial system cannot readily be mapped by most people. Modernism attempted to overcome the dilemma this posed for notions of representation. Distorted and symbolic forms of expression were used in an attempt in some way to map the otherwise unmappable. Late capitalism, with its 'unimaginable decentering of global capital', is said to have exacerbated the problem, going beyond the ability even of the modernists to provide any kind of adequate mapping.[31]

The inability to map socially is, for Jameson, 'as crippling to political experience as the analogous incapacity to map spatially is for urban experience. It follows that an aesthetic of cognitive mapping in this sense is an integral part of any socialist political project.'[32] Postmodern forms of representation are submerged in the landscape of multinational capitalism, able only to reproduce its own logic and incapable of gaining the critical distance necessary for an effective mapping. As a narrative which seems to dramatize something of the dilemma outlined by Jameson we might cite Troy Kennedy Martin's television series *Edge of Darkness* (1986), a nuclear conspiracy thriller within whose labyrinth the main protagonist is hopelessly disoriented. The issue is figured most explicitly when he gains entry to a high-security computer centre from which he manages to extract a map of the maze-like structures that offer illicit admission to the nuclear plant around which the plot revolves. A series of parallel mappings and labyrinths is evoked: those of the computer centre as bureaucratic stronghold, the database as electronic maze, the chart onto which are mapped the names of London underground stations, and the notorious public maze of the Barbican centre into which the protagonists escape from their pursuers in the computer building. Are we to believe that a mapping remains possible for those who have access to the information: a mapping both of the physical structures involved and of the conspiratorial web of knowledge? Or does the map of one maze only lead us into another? It turns out that what appeared to be radical discoveries and an unravelling of the conspiracy had in fact been pre-programmed, the protagonist duped and manipulated.

To resolve such difficulty we might seek to return to a clear distinction between map and territory, reality and representation, a reassertion of epistemological or ontological boundaries that seem to have become blurred along with their spatial counterparts. This would enable the observer to stand above the territory once more, as it were, to map it from an objective height rather than to remain implicated in it. This is rather like the experience of the protagonist of Rex Warner's novel *The Aerodrome* (1941). In the opening scene we find him wallowing, literally, in the soft and muddy marshland of his home ground. Such a sensuous existence, however, turns into disorientation when events throw into question the taken for granted realities of his family background and the traditional authorities of both church and the local squirearchy. At this point he becomes seduced into the attractions of the local aerodrome, eventually becoming a pilot and so seemingly able to free himself from implication by gaining a clear map-like view of all that lies below. It is towards such a view that Jameson might seem to aspire when he advocates the kind of totalizing discourse, in this case Marxist, so much criticized by theorists of the post-modern influenced by Jean-François Lyotard's *The Postmodern Condition* (1979). The aerodrome establishment of Warner's text turns out to be totalitarian in nature. It is wrong to suggest that anything other than an entirely fragmented Marxism should be equated with the oppressive qualities of the gulag. But we should question any theory or politics which seeks or claims to be able to map *everything* from its own lofty perspective.

It is not a matter of choosing simply between these two perspectives, the view from above or immersion at ground level. If the one produces a reductive or totalizing map, the other offers only another illusion. No existence is possible on unmapped ground. Cultural groups create their own mappings, imposing them on and writing them into the territory itself in a way that undermines the distinction between map and territory. Jameson's brand of Marxism is a map that precedes its territory, but so is any meaningful view. The kind of mapping he envisages may be as appealing as the prospect of understanding seemingly offered by the map of the labyrinth produced in *Edge of Darkness*, with all the ambiguities that involved. He is clear, however, that it would be something more complex than a simple act of mimetic representation, an 'as yet unimaginable' way of charting the world space of multinational capital.[33]

While in some respects suggestive, Jameson's historical account of the relationship between map and territory is unsatisfactory. We might in particular want to question how straightforward the relationship was between structure and experience, essence and appearance, in the initial phase of market capitalism. To what extent in that period – assuming that the three stages offer a useful outline structure – did most working people understand the kinds of relationships mapped out in Marx's *Capital*, for example? Any stage of apparent 'transparency', such as it was experienced, would have been historically-specific and the outcome of particular ideological effects at the time, rather than being the cultural norm. The outcome of what Jameson sees as a move away from this initial state is in fact more typical of the opaque cultural cartographies we inhabit. However much they might disagree on some points, Jameson and Baudrillard both put the argument in terms of the existence or not of an epochal change. The contemporary situation described in terms of the postmodern may have involved significant changes of one sort or another, and these might merit close study. But the distinctions between map and territory said to have been undermined in the postmodern era have always been arbitrary.

Extreme postmodern cases such as the production of virtual reality worlds may be suggestive of the way all social realities are mapped rather than exemplifying any kind of radical epistemological break. When we exit a computer-generated virtual world, do we step into an absolute reality or just another level of virtuality, albeit one reinforced by a network of powerful institutional forces? Visitors to Disneyland can experience a prototype virtuality in the 'Star Tours' feature. Strapped into their seats, they are propelled into the realistic illusion of being in a space vehicle, a screen at the front standing in for the window while stereo sound-effects and a lurching floor complete the effect as they seem to be inside the climactic scene of the film *Star Wars* (1977). When the ride is over, though, what is the world outside the simulator, the rest of the Disneyland enclave, other than another level of virtual reality? And what of the spaces beyond the boundaries of the theme park and into the world we usually understand as unquestionably real? Physiologically even, the particular shape of world that we experience as real is a model constructed by the brain from data supplied by the sense organs.[34] At the social level we are all constrained by the limits of our own cultural maps and the particular views of the world that they impose. Maps can be redrawn, but this is not always easy: unlike electronically simulated realities they cannot be changed at the throw of a switch.

Map and territory cannot ultimately be separated. Cultural mappings play a central role in establishing the territories we inhabit and experience as real, whatever their ontological status. The power to draw or redraw the map is a considerable one, involving as it does the power to define and what is or is not real. When a sharp distinction is made between map and territory a particular construction of reality is reified. A map that serves specific interests appears to be an objective representation of exterior reality. The mapped reality appears to be inviolate, existing on the territory itself rather than being the outcome of particular institutional and representative practices. Maps are granted the status only of passive representation. To blur the distinction between map and territory is to destabilize this relationship, to acknowledge the socially constructed character of the mappings within which our lives are oriented. It is also to create the possibility of change, although we should not underestimate the power with which particular mappings can continue to impose themselves even against our will. If in subsequent chapters we are drawn back repeatedly to the American experience it will not be because its culture represents the acme of the postmodern, as Baudrillard suggests, but because the powerful mapping of the reality of that 'fragile, mobile, superficial culture' offers a vivid illustration of a much more widespread cultural cartography.

2 World Views

> off the map: *out of existence; into (or in) oblivion or an insignificant position; of no account; obsolete* [. . .] on the map: *in an important or prominent position; of some account or importance; in existence* [. . .].
>
> *Oxford English Dictionary*[1]

Christian maps of the Middle Ages depicted the world as a flat disc centred on Jerusalem. Geography was subordinated to theology, while a variety of mythical beasts were shown to inhabit the unknown outer reaches. Maps compiled today with the aid of the latest developments in satellite and computer technology appear to offer a far more objective view, precisely delineating the outlines of the world. The difference between these two forms of mapping is only relative, however. Each gives a particular perspective on the world. Neither is entirely objective or entirely illusory. Like all other cultural products, maps are subject to strict conventions. What is represented and how depends on a number of contingencies as does the meaning attributed to the map. The map continues to owe as much to particular understandings of a territory as to the territory itself, if not more. There is and can be no such thing as a purely objective map, one that simply reproduces a pre-existing reality. Choices always have to be made about what to represent and how, and what to leave out. It is here that cartographic meaning is created. To be included on the map is to be granted the status of reality or importance. To be left off is to be denied. The reality that is given on the map is influenced by technical limitations and also by the deliberate strategies of the cartographer. More significant perhaps are the bounds set by the world-views of particular cultures, views that are themselves constructed and reinforced on the map.

Maps inevitably distort reality, as most cartographers concede. But the notion of distortion is misleading, suggesting as it does the possibility of some kind of pure, undistorted representation. A recurring dilemma resulting from technical limitations is the impossibility of reproducing exactly the curved surface of the globe on a flat sheet of paper. The difficulty is, as Jameson suggested, on the scale of Heisenberg's uncertainty principle in physics: to the extent that shape is maintained the representation of area is distorted, and vice versa. Various different projections have been used, some more or less useful than others depending on the priorities of cartographer and map user, but in no case can the problem be eradicated. Maps can also be oriented in different ways. We should note that this term, which has become sedimented in our language to refer to any kind of directional coordination, itself etymologically suggests a very specific 'orientation' found in Roman and medieval maps that put the east or orient, the direction of sunrise, at the top. We are so used to seeing the world with the north at the top of the map that we have difficulty recognizing places on maps that adopt any radically different orientation: we easily become disoriented. Yet our orientation towards the north is a relatively recent invention, coming with the introduction of the compass and its pull towards the magnetic pole. Other cultures have been oriented very differently. Arab cultures, looking towards Mecca, tend to have the south at the top; early American maps sometimes looked towards the west. There has been a widespread tendency to place at the top of the map the direction to which attention is most acutely turned.[2] An inverted world with the south at the top looks a very different place, as in McArthur's 'Universal Corrective Map of the World' (1979). Australia is at the top and centre with Europe relegated to the margins. Variations of orientation such as these are also to be found in other cultural frameworks. Each is, in one sense, arbitrary. There is no necessary direction in which we should orient ourselves. But it is equally true that, once mapped, such orientations impose themselves very strongly upon our lives: an elderly character in Keillor's *Lake Wobegone Days* dies from the disorientation brought on merely by moving to a house facing a different way. Cultures or nations also tend to put themselves at the centre of their mapped worlds, often at an exaggerated scale. Early Christian maps are centred on Jerusalem; Rome, unsurprisingly, was at the centre of maps of its empire; ancient Chinese maps had China at the heart of the world; and so on. Britain achieved similar grandeur in maps of its empire, formalized with the adoption of Greenwich as the prime meridian in 1884. A comic evocation of this tendency is found in Daniel Wallingford's maps of the United States, drawn according to the perspectives of the occupants of New York and Boston. In each case the home city is shown in huge disproportion, accounting for much of the east coast, while the mapping of the rest of the nation is blurred and stereotypical.[3]

Difficulties arising from choice of orientation or projection are not generally seen as challenging the notion that the map remains in the end the product of the territory, rather than something that might precede or overlay it like the maps of Borges, Carroll or Carey. Cartographical theorists tend to adopt a standard information theory approach. The map is seen as a communication channel for information to be transmitted from one place to another. Data are to be taken from the real world before being encoded in map form and then decoded by the map user.[4] Cartographers who adopt this approach do not deny the possibility of distortion. It is treated as 'noise', a disturbance at some point in the communication channel. The importance of this noise is acknowledged; keeping it to a minimum is seen as the principal task of the cartographer. But this approach presupposes that noise is essentially secondary, intervening between an otherwise smooth translation of the real world into mapped image. The possibility that the map may affect understandings of the real world itself, radically destabilizing such a schema, is not usually entertained. Where it is, the effect is put down to error or abuse. As one writer puts it:

> The map is only reflecting the mutable world in a passive way; it is not a living, active thing. Although it is true that in extreme cases information which has been erroneously recorded on maps has actually incited people to change the geography to conform to the map, such rare occurrences are contradictions of map reading principles.[5]

But the map is not so passive, nor merely a reflection. Such extreme cases may be only the tip of an iceberg treacherous to this author's project. Maps can be very persuasive, in numerous ways. They can construct relationships, unities and divisions that have important implications, for example in international relations. A map can organize the world according to almost any principle of order.[6] It offers an instant picture of the whole that often appears natural and objective rather than an expression of ideology. The same case put in words might not so easily conceal its roots. The best way to argue against the world-view expressed on one map is to offer a rival projection.

Maps have the power to shape our view of the world as well as to reflect it. Some of the political implications are suggested in Michael Kidron and Ronald Segal's *The New State of the World Atlas* (1991), in which a range of maps and cartograms are used to highlight the relationships between nations in terms of indicators such as population, debt, military spending, human rights and the availability of food. A similar approach is found in the *Strategic Atlas* (1983) compiled by Gérard Chaliand and Jean-Pierre Rageau. A number of projections of the world are shown in which international relations seem to take on different shapes. The map of the world that has become dominant in the West is a Eurocentric view in which Europe and Africa are flanked on either side by the Americas and Asia/Australasia. A map that places the Americas in the centre has different implications: the United States appears much closer to its old superpower rival, the Soviet Union, while the Pacific basin becomes an area of central importance rather than being confined to the cartographic margin. Circular or polar projections further defamiliarize our view of the world, both geographically and politically. Foreign policy strategies are affected by these perspectives. According to different readings of the map, geopoliticians have developed general theories seeking to outline fundamental dynamics underlying world history and struggle. An influential view was presented by Halford Mackinder in 1904. Central Eurasia was designated as the global 'Heartland' around which control of the planet revolved. On Mackinder's map (subjected to later revisions) the Heartland lay at the centre of a structure completed by a 'marginal' internal ring, stretching from Western Europe and the Middle East to Japan, and a further ring of islands or 'outer' continents, a hierarchical perspective that would not have appeared so neat or convincing if applied to any other than the Eurocentric projection (both cultural and cartographic) onto which it was drawn.

Maps such as these have a rhetorical character that may be all the more potent for not generally being recognized. Maps have been used widely by writers of fiction to make their worlds more real. Their authoritative appearance can also be used for political ends. As one writer puts it: 'Map-conscious people

[. . .] usually accept subconsciously and uncritically the ideas that are suggested to them by maps.'[7] We tend to trust maps or to take them for granted.[8] A map can give a particular slant to an existing state of affairs. It can also be used to argue for or against change. As Hans Speier puts it, maps 'may give information, but they may also plead.'[9] Maps have often been used as instruments of propaganda. Nazi Germany produced a range of propaganda maps during the Second World War. In one example two maps were juxtaposed, one showing Germany at the top of an otherwise empty rectangle, the other depicting Britain along with its extensive colonial haul: which was really the aggressor, it asked pointedly.[10] More recently, the Labour Party resorted to cartography in a poster campaign designed to make the most of Conservative divisions over the strengthening of links between Britain and the rest of Europe. Moved on the map a few thousand miles to the west, Britain was depicted as sitting alone in mid-Atlantic, symbolizing the isolationist position of those on the Tory right.

Political boundaries often take on a particular significance on the map. 'A map on which boundary lines are changed or deleted may constitute an annexation of territory on the symbolic plane.'[11] Makers of atlases often face difficulties in the representation of disputed boundaries, so important is their status. A number of different and confusing conventions have resulted, such as marking disputed boundaries with dotted lines. But this only raises further difficulties. Who is to decide which borders are in question, and what is their basis in the first place? Some atlases attempt to evade the problem by claiming that the boundaries they show do not imply any principles of international recognition.[12] But such disclaimers do little to reduce the power of the map to strengthen territorial claims. A despairing solution was attempted on one map produced by British European Airways: all political boundaries were omitted on the sheet covering continental Europe to avoid causing offence to any potential traveller.[13] Kidron and Segal offer a map in which the number of states depicted as having no unresolved boundary disputes is reduced to a handful.[14]

These difficulties cannot ultimately be contained. It is not a question simply of the deliberate distortions of propagandists, however effective these may be. Maps affect our understanding of the world at a more subtle and fundamental level. When the first printed maps appeared in Europe in the 1470s, for example, they began a process of profound change in people's experience of space. Most had lived out their lives with only a vague notion of the world beyond the bounds of their daily experience. Maps first came into widespread use in England in the Tudor period. They provided instruments of control for landlords and governments as well as influencing conceptions of place at various geographical scales.[15] With the aid of the map, territory could be understood as a whole rather than as a series of separate local impressions. The possibility was born of knowing distant places, although these often remained rooted as much in the imagination as in the real world.[16]

Maps can be read on a number of levels, like other cultural or artistic products. Focusing mainly on technical improvements, some historians of cartography tend to exaggerate the difference between maps and other forms of representation, 'even suggesting that in the Renaissance what is seen as a split between cartography and art was finally accomplished through the introduction of more technically perfect methods of surveying and map drawing.'[17] The early history of maps is largely inseparable from developments in the world of art. Map-making and landscape painting were often the work of the same artists, including luminaries such as Leonardo da Vinci. Maps have long been used for decoration and the demonstration of artistic skill. Until the Renaissance no terminological distinction was made between painting and map.[18] The split that was implied at that time was itself an arbitrary line imposed on the map, as will be seen below. All maps remain the products of particular representative practices. Three different levels of cartographic meaning might be distinguished: the immediate practical use of the map, its symbolic meanings and their social implications.[19] Practically, for example, Tudor estate plans aided the administration of the estate and developments such as the enclosure of the land; symbolically, they signified proprietorship and authority; socially, this played a part in the maintenance of a class structure based on the private ownership of land.[20]

Maps have served a great many practical purposes. From the stick charts of the Marshall Islanders to charts developed by sixteenth-century explorers and modern road and rail maps, they have always been used for navigation. Maps have also been an important feature of military campaigns. They were used by warring states in ancient China. The Ordnance Survey in Britain, as the name suggests, owes its origins to military uses. Napoleon Bonaparte recognized the importance of maps to his conquests. So did his opponents. The most frequent subjects of early mappings in Britain were areas of military sensitivity. The first scientific survey was carried out by the military in southern England amid fears of attack after the French Revolution. Cartographers repeatedly turned their attention to vulnerable areas such as Scotland, Ireland and the Channel coast. Popinjay's early map of Portsmouth (*c.* 1584) showed defensive works proposed to meet a threat of Spanish attack. While some parts of the country remained uncharted, struggles between the European colonial powers led to the ports of Plymouth and Portsmouth having permanently established Board of Ordnance drawing offices to prepare fortification plans.[21] By 1800, Portsmouth had been mapped more often than almost any place of comparable size in the world.

The subjugation of Ulster from 1593 was recorded on a series of at least eighteen maps by Richard Bartlett, an English map-maker employed by the Irish Lord Deputy. The mapping of Ireland was carried out on a piecemeal basis to answer the needs of a succession of military and political developments, a process that culminated in the more systematic Ordnance Survey mapping authorized in 1824.[22] Another mapping rooted in conflict was the basis of the 1921 settlement which drew the arbitrary line across the territory on which the current struggle is founded. In Scotland, Cumberland demanded maps to help pacify the Highlands after the crushing of the Jacobite rebellion at Culloden in 1746. The task went to Major-General William Roy, who, emphasizing its vulnerability to invasion, advocated the mapping of the whole of Britain, an enterprise eventually begun in 1791 by the Board of Ordnance although not completed until 1870.[23]

In the early nineteenth century more rapid lithographic techniques enabled maps to be provided within ten days of military actions. A mobile printing press was used in the field by Wellington. A number of developments in mapping techniques came in response to shortcomings demonstrated at great expense to both sides in the American Civil War.[24] By the time of the two world wars, large print runs of battlefield maps were available within a day and a half.[25] The continuing need for accurate maps in wartime was underlined during the US invasion of Grenada in 1983. Troops were supplied with only hastily printed copies of outdated British charts and a tourist map onto which a military grid was superimposed. An air attack led to the destruction of a mental hospital that was not marked on the maps in use. One US soldier was killed and another eighteen were injured by another attack that was ordered by an officer using one set of grid coordinates but carried out by aircraft equipped with a different set.[26]

The importance of maps is not restricted to these immediate uses. Maps might be needed for the defence or administration of the nation, but they play an equally important role in the creation and sustenance of the very idea of the nation state.

> The map is the perfect symbol of the state. If your grand duchy or tribal area seems tired, run-down, and frayed at the edges, simply take a sheet of paper, plot some cities, roads and physical features, draw a heavy, distinct boundary around as much territory as you dare claim, colour it in, add a name – perhaps reinforced with the impressive prefix of 'Republic of' – and presto: you are now the leader of a new sovereign, autonomous country. Should anyone doubt it, merely point to the map. Not only is your new state on paper, it's on a map, so it must be real.[27]

The national map and the concept of nationalism are inextricably linked, the map doing more than expressing a pre-existing sense of nation. The drawing of the national map can help to create nationalist sentiment. The growth of nationalism might also benefit the map-maker. Early articulations of national consciousness gave fresh impetus to the development of mapping techniques in Europe. The results were seen in the production of atlases of England and France in the late sixteenth century. Queen Elizabeth commissioned Christopher Saxton to survey England and Wales and to publish the maps in atlas form. His maps, in turn, helped to establish the territory: 'however faithfully they may have gathered and repeated the "facts" of England's history and geography [Saxton and his contemporaries] had an inescapable part in creating the cultural entity they pretended only to represent.'[28] A few years later Henry IV of France commissioned Maurice Bouguereau to draw up a similar work to mark symbolically the recent unification of his country.[29] The national atlas was revived as a potent symbol of nationhood after the 1939–45 war when the format was used by a number of newly independent states. The new states were merely turning to their own advantage cartographic techniques that had been used repeatedly by colonial powers to legitimate their territorial claims.

Nations often produce maps which claim as their own parts of neighbouring lands, a dramatic example being a billboard-sized map of Guatemala mounted at its border with Belize: the latter simply does not exist on a chart that shows Guatemala stretching uninterrupted to the Caribbean, despite its formal recognition of Belize in 1991. Maps can also assert more distant claims. Maps were used extensively in the coverage of the war between Britain and Argentina over the Falklands/Malvinas in 1981, although even the most creative of map-makers would be pressed to make a cartographic case for the British claim to the islands. Relatively few maps showed the real relationship between Britain and the distant islands in the south Atlantic, most remaining confined to the southern hemisphere. Where the true relation between Britain and the islands was shown it was likely to be in a work arguing against the British claim. Such maps could still serve the ideological project of the Thatcher government, however. What was asserted was less a claim to legitimacy than a demonstration of a continuing ability to conduct military operations at a distance. The fact that the islands were so far away, off the edge of the map, only emphasized the achievement. Britain might not have been present on many of the maps, but its power was represented by dynamic arrows sweeping over the horizon to illustrate the tentacles of supposed imperial might.[30] It was a classic reassertion of the myth of British 'greatness', looking back to the days of a grand empire based on naval dominance at a time when the reality for those at home was one of gloom and recession.

World maps and globes have helped to achieve and justify colonial conquest. Their importance also goes beyond immediate practical purposes of navigation or defence. The blanks left on maps were a tantalizing invitation to the would-be explorer or colonist. More sophisticated maps whose grids brought the entire globe into what appeared to be a single rational framework played a part in the process, as I will suggest at greater length in later chapters. The production of maps of colonies is a symbolic form of conquest. To map a territory is to stake various kinds of claim to it, to make assertions of ownership, sovereignty and legitimacy of rule. The map may also increase the ability of the imperial power to generate wealth from the colony, as in the case of the great surveys of India conducted on behalf of the British East India Company in the eighteenth and nineteenth centuries. Maps can also be used to contest imperial claims or to demonstrate the realities of colonial rule; showing, for example, how the communications networks constructed by Western powers permitted the exploitation of natural resources and their removal from the country rather than enabling them to benefit their own people.

Mapped demarcations were used by the rulers of Spain, Portugal and Britain during their competing expansion programmes in the Americas from the sixteenth century. One of the earliest known existing maps to show the New World, the 'Portolan World Chart' (*c.* 1500) attributed to Juan de la Cosa, staked a claim for the British. The map declares that John Cabot – the Italian navigator Giovanni Caboto sailing under patent for Henry VII – discovered North America in 1497.[31] Rival claims from Portugal were accorded due cartographic glory in the 'Cantino Planisphere' of 1502 commissioned by Albert Cantino, envoy of a member of the Portuguese court and based on the findings of Gaspar Corte Real, who returned to Lisbon from the New World the previous year. Here the land is named Terra del Rey de Portuguall. It is placed, conveniently, east of its actual location to lie on the Portuguese side of the arbitrary line dividing the non-Christian world into the two spheres accorded to Portugal and Spain by a Papal bull issued in 1493.[32] This division, formalized in the 1494 Treaty of Tordesillas, is a classic instance of the purely abstract line on the map that comes to dictate the terms of subsequent history.

Rival British and French claims were also asserted on the map as well as the territory. Guillaume Delisle's 1718 'Carte de la Louisiane et du Cours du Mississippi', in addition to remaining an influential chart of the Mississippi, was a 'chauvinistic pictorial argument supporting French authority [that] showed the British colonies surrounding the French possessions in the West, and claimed the Carolinas for France.'[33] The British reacted, citing as evidence Herman Moll's 1715 map, 'A New and Exact Map of the Dominions of the King of Great Britain on ye Continent of North America'.[34] John Mitchell's 'Map of the British and French Dominions in North America with Roads, Distances, Limits, and Extent of the Settlements' (1755) was another deliberate intervention in the ongoing cartographic dispute. Described as 'the primary political treaty map in American history', it continued to exert influence for many years. The map was used in American boundary disputes as late as those between Canada and Labrador in 1926, Wisconsin and Michigan in the same year, and between Delaware and New Jersey in 1932.[35] A line drawn from north to south on the British imperial map according to a Royal Proclamation of 1763 marked the point at which settlers were told they could not go in the search for new land. Intended to meet the trading needs of the mother country by concentrating settlement along the seaboard, the demarcation was bitterly resented, contributing to the accumulation of colonial grievances that led eventually to the revolutionary break from British rule.

If boundary lines such as these, both between and within the territories of rival colonial powers, were a key aspect of the mapping of America, another was the act of fixing names onto the map, a calling into being of places that also inscribes a dynamic of appropriation. When Columbus arrived in the 'Indies' he was well aware that the natives already had their own names for the islands, but this did not stop him superimposing his own. The islands themselves, capes, mountains, points and ports were all confidently assigned Western names in a ritual of conquest, an act of conceptual appropriation seemingly inseparable from the seizure of the land itself. What was involved was an assumption of sovereignty equal to that with which Adam named the animals in Eden. From being a strange, incomprehensive land whose peoples spoke unfamiliar languages, the New World was converted into a legible text from which a colonial history could flow. Juan de la Cosa's map wrote onto the continent a series of names of his own invention that failed to stick or to have any discernable influence on future mappings. The ability to impose names that would take root in the territory rather than remaining a superficial and idiosyncratic tracing was to become a significant indicator of the success of the colonial venture.

That America was itself so named was almost entirely due to the decision of the cartographer Martin Waldseemüller to inscribe the name, after the explorer Amerigo Vespucci, across the southern half of the continent on his 1507 extension of the Ptolemaic world map. Whether or not Vespucci deserved such an honour remains the subject of controversy. Having championed Vespucci in 1507, according him equal prominence with Ptolemy in the decorative border above the chart, Waldseemüller seems to have had his own doubts, both about the explorer's claims and the shape of the continent attributed to his findings. On subsequent maps of 1513 and 1516 Amerigo Vespucci's name was conspicuous by its absence, replaced by more familiar legends such as 'Terra Incognita' in works that retreated from the earlier map's bold assertion that the territory was a New World distinct from the Asian mainland. Waldseemüller may have come to doubt Vespucci's claims and in doing so dismissed what we would consider to be by far the age's most 'accurate' rendition of the shape of the New World. Vespucci's reputation continued to be uncertain. The major writings published under his name at the time are now widely believed to have been crude forgeries, spiced-up narratives containing absurdities that made him appear to have been an arrogant liar. They were influential, however, and provided some of the first widely circulated images of the inhabitants of the New World.

Vespucci's glorification on the map might have been merited less by his being the first European of his age to see the New World, as one account claimed, than for the act of interpretation that enabled him to see it as a new continent rather than just an extension of Asia. Quite what either he or Columbus believed at any one time, however, remains uncertain. Whatever the reason, Waldseemüller's map of 1513 credited Columbus with the discovery of the continent. It may already have been too late, however, for the alternative name 'Columbia' to have been fixed upon the territory. Mercator's world map of 1538 restored the notion of a new landmass distinctly separate from Asia, granting the name 'America' for the first time to both the north and south of the landmass. The name stuck, proceeding in history to gain enormous symbolic resonance around the world, what it signified depending on whether you were the victim or supporter of its owner. For Garrison Keillor a more deserving figure might have been Eric the Red, a leader of the Viking settlers thought to have reached America some 500 years before the explorations of Columbus or Vespucci. The continent should thus have been called 'Erica', a name that would have given a distinctly different tone to those American institutions: 'The United States of Erica. Erica the Beautiful. The Erican League.'[36] The great investment so many of its people have in the name 'America', as a totemic mark of identification in a highly diverse and socially divided nation, is a demonstration of the potency of such arbitrary cultural constructions.

A similar attempt to achieve conceptual as well as military control was inherent in the English colonial mappings of Scotland and Ireland. Under the cloak of rationalization and enlightenment, the names of places and geographical features were changed, depriving the population to some extent of their own legacy of meaning, a process dramatized in Brian Friel's play *Translations* (1980). In some cases place names were simply Anglicized, as part of the enterprise of colonial reinscription and domination. But a more subtle process was also at work. Ancient Irish names and boundaries were also written onto the nineteenth-century six-inch Ordnance Survey map: apparently more authentic forms that could serve to legitimate a mapping that was located in the colonial present. 'Irish place names were not displaced but co-opted. They were a major feature of the map but they bore no witness to Irish ownership of the soil. They endorsed, rather, two political facts: the state of Union between Britain and Ireland and the power of the English Protestant landowners, the Ascendency class, in Ireland.'[37]

Cartography has also been employed in a wider realm of cultural discourse than that involved in the claims of nation or empire. Maps have always been used to construct or support particular ideological views of the world, in ways that go well beyond the bounds of simple propaganda. This is neither a new phenomenon nor one that is absent from modern maps. A surviving ancient Egyptian map carved into a wooden sarcophagus depicts for the deceased a route to Paradise based on the valley of the Nile. This is an early example of what one commentator terms 'a *theoretical map* or model, a persistent form in cartography in spite of the admonitions of some that geographers should address themselves only to the real world.'[38]

Christian theology accounted for the shape and features of the world in almost all the early maps produced in the West. The *Christian Topography* of the monk Cosmas Indicopleustes, written between AD 535 and 547, set itself to demolish false doctrines about the universe and to replace them with a true picture. Fancy notions of a round world propagated by the satanic errors of the Greeks were dismissed by an appeal to common sense. An image was presented of a flat earth vaulted by the arch of heaven. Evidence supporting this supposedly all-encompassing geography was drawn from scriptural texts such as St Paul's observation that the tabernacle of Moses was designed to be a figure of the world in its furnishing and layout.[39] It would be wrong to assume that this was the work of mere ignorant superstition. Cosmas had travelled extensively and dabbled in astronomy. His world view is better seen as an effective communication of elements of the religious doctrine of his time. The early maps of the Middle Ages adopted a similarly theological framework. The 'T–O' maps, as they have become known, reduced the shape of the world to a 'T' structure contained by the 'O'-shape boundary of the world. The horizontal stroke of the 'T' was constituted by the meridian running from the river Don to the Nile while the vertical axis was provided by the line of the Mediterranean in a map 'oriented', with the east at the top.

Maps such as these clearly put the expression of an ideology before any attempt to establish what today would be claimed to be a literally accurate representation. The term 'accuracy' in this use is misleadingly teleological, however, implying that what is charted by the history of map-making is nothing more than an evolutionary process of improvement. A better notion might be one of efficiency.[40] The maps of the Middle Ages were regularly used as metaphors to express abstract ideas. However inaccurate they may seem to us, they were efficient vehicles for the transmission of certain world-views. They were accurate charts of the beliefs of their time. Maps accompanying encyclopedic histories included features such as the location of the Garden of Eden and the landing of Noah's Ark.

The cartographic image became 'a multivalent symbol capable of expressing a host of different moral and religious meanings [. . .].'[41] The basic 'T–O image' was retained in medieval maps that became gradually more sophisticated. The Psalter *mappa mundi* in the British Museum, for example, remains circular in shape but the 'T' is more complicated, the outline of the land more coast-like, the seas including inlets and looking a little more like seas than wholly abstract symbols. The theological message continues to dominate, however; the figure of Jesus is at the top of the map, overlooking the world from on high with angels at his side.

The famous Hereford map, created from 1280 to 1300 by Richard de Bello, priest of Haldingham in Lincolnshire and later of Hereford cathedral, includes a number of clearly recognizable geographical forms. The names of many European cities are included, alongside biblical scenes such as Moses on Sinai and Noah's Ark. The map also includes a wealth of bizarre mythical figures and beasts around the edges of its representation of the known world. There are humanoid figures with no heads, with eyes in chests or with arms growing out of heads. The material is drawn from a variety of sources including narratives like those of Julius Solinus in the third or fourth century AD, accounts that themselves were poached from earlier sources such as Pliny and were greatly to influence the future shape of Christian geography.[42] Much of the detail on the Hereford map was copied from copies of other materials that were themselves works of invention. Not that this was to prevent the moral panic surrounding the map in 1988 being conducted in terms of its authenticity and value as a precious original. One of the characteristic features of Western obsessions with notions of the authentic is a reification that fixes cultural products at certain points in what is really an ongoing process of simulation.

What might seem a strange combination of the geographical, the biblical and the pagan on the Hereford map is in fact a revealing chart of the heady mixture underlying medieval belief. The contemporary Ebstorf map, believed to have been drawn by an English priest, Gervase of Tilbury, and destroyed in the Second World War, was similar in general design and layout, although the T–O framework was further complicated by the addition of more watercourses that break up the masses of land. The Ebstorf map was intended by its author to be of practical geographic use, an inscription offering 'directions for travellers and things on the way that most pleasantly delight the eye'. Such pragmatic concerns are far from eclipsing the theological dimension: the entire map is depicted as lying in the embrace of a crucified Christ who marks out the points of the compass, his head appearing at the top of the image with the feet at the bottom and a hand protruding at each side.

Religious frameworks are not the only ones within which maps of the world have been constructed. An extraordinary map of 1574 produced in Cologne by Georg Braun depicts the continents in three sections unfolded beneath the wings of the double-headed eagle of the Holy Roman Emperor, then Maximilian II. Less imperiously, a French map produced in the following year encompassed the world within the frame of the head of a jester. In a more local example of self-aggrandizement, a German map of 1581 charted the three continents of Europe, Africa and Asia in the pattern of a clover-leaf, the emblem of Hanover. The shape of a rampant lion was used for a series of maps produced by supporters of a united Netherlands before and after the split between north and south was formalized in 1648; another lion, facing the other way, was used to depict the separate Dutch state.[43]

It would be wrong to suggest that the expression of symbolic meanings, religious or otherwise, simply disappeared with scientific improvements in cartographic technique:

> Indeed, the realism of the image may have served to strengthen the force of symbolic as well as literal meanings. An example of the latter is the way the new scientific celestial cartography of the Renaissance continued to be harnessed for astrological purposes in Tudor England. [. . .] Although astrology perpetuated 'unscientific' ideas, it nevertheless helped to popularise the new science, such as Copernican conclusions about the universe; moreover, the revival of English astrology in the second half of the sixteenth century was also closely linked with the rise of the applied mathematical sciences.[44]

Maps produced by Opicinus de Canistri in the 1330s drew upon the Italian and Catalan portolan charts used for navigation at sea. Unlike much of the cartography of the period, these were based largely on direct observation by means of the newly developed mariner's compass. They were the most advanced and 'accurate' maps of their time. Yet they were also used in this case to illustrate moral issues. The coast of north Africa on one map is shaped as a woman, that of the northern Mediterranean as a man, his right ear and her nose framing the strait of Gibraltar in a map seen as symbolizing the sin of mankind.[45]

Great strides were made in the science of mapping in the fifteenth and sixteenth centuries, but symbolic additions thrived unabated. Fra Mauro's *mappa mundi* of 1459–60 is highly detailed, including the mapping of roads reported by contemporary travellers. In terms of the weight of detail and information it represents great progress from its predecessors. But much of the information is either superfluous or spurious: 'The geographical foundation, rational as it is by fifteenth-century standards, has been overlaid with a swarm of inscriptions rehearsing geographical information (distances, climates, natural obstacles), political circumstances, historical, biblical, and classical events, and tales of mythical lands, and all the other baggage of medieval *mappaemundi.*'[46]

Jacopo de Barbari's celebrated *View of Venice* of 1500 was considered a great technical feat, which it no doubt remains. But there was more to it than an apparently objective rendition of the city. The incredible detail has a rhetorical effect, tending to convince the viewer that it is literally accurate. The map is in fact inaccurate in many details, constantly reverting to conventional forms rather than objective representation. It also gives exaggerated emphasis to central functional and symbolic nodes of the life of the city: the ceremonial and administrative area of San Marco, the Rialto commercial and financial centre, and the military-industrial focus of the Arsenal.[47] The primary meaning of the map is metaphorical. Above the view is depicted Mercury, the patron of commerce; accompanying text says the god shines favourably on the city. Below is Neptune, lord of the seas, who is said to smooth the waters of the port. These are not mere decorative additions to the image but integral parts of its meaning, as Juergen Schulz suggests. The city is celebrated as the principal trading and maritime power of Europe at the time: 'Jacopo's print is a visual metaphor for the Venetian state, like the familiar symbols of Venice, the winged lion of Saint Mark, and the female figure holding the scales and sword of Justice which represent the divinely protected and justly governed Republic.'[48]

The combination of faith and reason in Barbari's depiction of Venice is not contradictory; the two existed side by side in the spirit of Renaissance humanism. By partially idealizing a real rather than a purely imaginary city, Barbari's map 'celebrates a myth of Venice as having already achieved that perfection to which humanist writers and architects elsewhere only aspired.'[49] Rather than gaining in objective accuracy, as a rational teleology might expect, later perspectives of Venice, modelled on Barbari's, showed a growing willingness to put symbolic matters before realism: 'In these maps the lido and the coastline are increasingly distorted toward a circular [ideal] shape, the urban area compressed to a more compact than linear form and in some cases the circular map is segmented by orthogonals focusing on the Piazza San Marco, the centre of the city.'[50]

Other maps of the time reveal similar efforts to reconcile apparently contradictory discourses. Also produced in Venice, in 1485, was a world map by an anonymous cartographer that attempted to bring together in the same work Christian concepts such as the existence of Paradise in the east as the source of the four main rivers of the world and the findings of both Ptolemy and the latest voyages of discovery. Regardless of such developments, Antonio Saliba's map of 1582 reverts to the verities of medieval cosmology, presenting the old view of the world as a series of concentric circles into two of which are sketched a map of the continents. More up to date in its general form, a map produced around 1597 by Jodocus Hondius, one of the great Dutch cartographers, superimposes upon a huge southern continent a series of allegorical figures the theme of which is elaborated by verses from the Bible. A Christian knight does battle with the assorted evils of the world pictured as figures of Sin, the Devil and Death. The map thus dramatizes a fear of the unknown threats seemingly posed by the new areas of the globe being opened to exploration at the time. Onto such regions is projected a duel between the forces of 'good' and 'evil' that would prove an effective rationalization for colonial conquest. A closer look at the detail suggests that the map also represents the contemporary struggle between Protestant and Catholic.[51]

The decorative content of maps, such as that found on the Hondius map or Barbari's view, has been neglected by many cartographic historians. Decoration is often viewed merely as superficial embellishment. It tends to be ignored or treated as no more than a marginal question of little relevance to the serious business of the map. Decorative detail can play an important part in the meaning of the map, however, serving a variety of moral, religious or political ends just as effectively as the rest of the image.[52] The decoration of the map can have important ideological significance. On Christopher Saxton's general map of England and Wales in his atlas of 1579 the rich decoration is part of the celebration and deliberate mystification of a golden age of English empire. The prominence of the royal coat of arms on Saxton's maps symbolized the monarch's sovereignty over the land. It has also been suggested that the map might at the same time have undercut monarchal authority by strengthening 'a sense of both local and national identity at the expense of an identity based on dynastic loyalty'.[53] According to this reading, it was the actual position of the monarch that was shown to be superfluous decoration: 'Maps thus opened up a conceptual gap between the land and its ruler, a gap that would eventually span battlefields.'[54]

None of this is a question of the errors and fancies of earlier times against which we can confidently set the objectivity or accuracy of modern maps. The history of mapping does not follow a simple line of increasing objectivity. Italian cartographic supremacy, for example, was lost to the Dutch in the late sixteenth century. Mapping played an important part in the great colonial explorations of the Dutch, yet there was if anything an increase in decorative detail which led in some cases to a reduction in technical accuracy.[55] French maps of the seventeenth and eighteenth centuries restored principles of reason, symmetry and proportion, but this was no more than the imposition of another cultural perspective. In none of these cases, nor in modern maps compiled with the aid of the latest technology, should cartographical representations be seen as neutral phenomena. All maps impose their own particular realities onto the world. Even the humble road map is more than just a practical aid to navigation. Highways are the principal features of the landscape on the road map or atlas, a dominant form of mapping that reflects the rule of the automobile in many Western societies. The road map traces its own preferred meanings onto the territory, supporting a mythology of individualism symbolized by the supposed freedom of the open road. Social as well as geographical mobility is implied, especially on maps of the American highway network. Rigid structures of race, sex and class are effaced. Alternative, perhaps more collective, means of negotiating the territory may be denied. Apparently a celebration of rootless individualism, maps of the American interstate network built with federal dollars under a congressional act of 1956 are testament instead to the corporate lobbying power of automobile, oil and construction interests.

The ideological dimensions of one map may be revealed by another. A classic case is the assault on the Mercator projection of the world undertaken by Arno Peters since the 1970s. The Mercator projection, based on the work of the Dutch cartographer Gerardus Mercator (1512–1594), is the only one in which the lines of the compass directions appear straight on a flat surface. For this reason it was widely adopted for navigation. Variants of the projection were for years almost undisputed as the standard framework for maps of the world. Yet it suffers from serious distortion. Areas of extreme latitude are greatly exaggerated in size. Almost all of the land affected by this exaggeration is in the northern hemisphere, the far south containing relatively little land mass. Northern lands thus appear to account for a far greater proportion of the surface area of the world than is really the case. The effect is 'the equivalent of looking at Europe and North America through a magnifying glass and then surveying the rest of the world through the wrong end of a telescope.'[56] Greenland appears to be larger than China, although it is less than a quarter of its size. Scandinavia is shown the same size as India when on the ground India is three times as big. Europe appears to be larger than South America, although the latter is nearly twice its size. The so-called Third World is significantly reduced in size. It may thus appear less important, a tendency liable to support the perspective of colonial or neo-colonial powers in the northern hemisphere.

The Peters projection, similar to one described by the Scottish clergyman James Gall in 1855,[57] has challenged Mercator's view. Peters describes the Mercator atlas as 'the embodiment of Europe's geographical conception of the world in an age of colonialism'.[58] His projection represents each country equally in terms of its area. All countries and continents are presented on the same scale, their relative sizes maintained although at the cost of inevitable distortion in shape. For Peters, 'this equal presentation is the expression of a consciousness that is gradually replacing our conventional way of thinking about the world.' This is not to say that the Peters projection is a neutral or objective intervention. Far from it. It is as much a political as a geographical tool. Peters launched his view of the world amid much publicity and with an attack on the Mercator projection that was welcomed by a number of United Nations and other international bodies. Free copies of the Peters projection are distributed, for example, to subscribers to the magazine *New Internationalist.* Mark Monmonier argues that organizations such as UNESCO might have had excessive faith in the projection. A demographic cartogram showing the size of each country relative to its population might better have conveyed their argument about the imbalance between resources and needs. Cartograms such as these show clearly how maps can sometimes best communicate data through distortion rather than the pursuit of literal accuracy. This is a tradition in travel maps that goes back as far as the Peutinger Table, a Roman route map presented in a long strip format covering a thousand miles from Britain to India. Perhaps the most famous modern example is Henry Beck's stylized map of the London underground. The map it replaced was more accurate but was also complex and confusing. Beck's map is simple, elegant and easy to use, although it can give the visitor a strange sense of London geography: the formalized view from beneath the streets can seem to override any real sense of direction on the cluttered surface. Where the map of the world is concerned such blatant stylization lacks the rhetorical power of conventional projections, even those as unfamiliar as Peters', which are usually reluctant to forfeit a claim to represent the world more objectively.

Whatever else it does, the Peters projection retains the same orientation as Mercator. The 'advanced' north may appear shrunken in size, but it remains physically above the south, in a position historically far from accidental. That which is physically above is still liable to be considered superior in the other sense of the term. Such meanings are conveyed on maps in a powerful yet implicit manner to which attention is rarely drawn in daily life; to invert the projection is to turn the world upside-down in more ways than one.

3 Maps of Meaning

The fundamental codes of a culture – those governing its language, its schemas of perception, its exchanges, its techniques, its values, the hierarchy of its practices – establish for every man [sic], *from the very first, the empirical orders with which he will be dealing and within which he will be at home.*

Michel Foucault[1]

Maps enable us to gain a sense of our place in the world, to orient ourselves. In this way they are like all socio-cultural frameworks: 'cultures are *maps of meaning* through which the world is made intelligible.'[2] As Edmund Leach suggests: 'Our whole social environment is map-like. Whenever human beings construct a dwelling or lay out a settlement they do so in a geometrically ordered way. This seems to be as "natural" to Man [*sic*] as his capacity for language. We need order in our surroundings.'[3] Without maps, life can be difficult and even dangerous. The protagonists of Ian McEwan's novel *The Comfort of Strangers* (1982), for example, find their lives under threat while on holiday in Venice after venturing out at night and getting lost without their tourist maps. The text implies that they are also without the appropriate cultural maps, failing to pick up clues or to recognize the approach of danger in the form of the characters Robert and Caroline who cross their path. In the film adaptation we see a reproduction of Jacopo de Barbari's map hanging in Robert and Caroline's apartment, symbolizing perhaps their greater control of the territory. Robert's attitudes towards sexuality – both his sadistic practices and his belief that women should be kept in 'their place' – certainly demonstrate such a concern for rigid control at the cultural level.

We are all to some extent like the occupants of Peter Carey's imaginary land of the cartographers, often craving the kind of security that comes from knowing where we are, both socially and territorially: 'So we make maps of social space by using territorial space as a model. In such usages the more featureless the context of actual territorial place, the more rigid and artificial the model has to be.'[4] Leach cites the example of tent dwellers living in barren and naturally featureless plains in

Mongolia who observed rigorous rituals in the controlled space of the tent as if to make up for the absence of clear points of topographical orientation. Cultural maps compensate for a lack of any strongly directed human instinctual drive. The worlds we inhabit are largely cultural rather than natural and as such are subject to a wide variety of cartographies. None of these is fixed on any ultimate or transcendent ground. Their provisionality may, paradoxically, explain the firmness with which they are often imposed.

Some argue that universal or innate maps can be identified, usually at the level of the physiological structure of the mind. For Claude Lévi-Strauss, 'the unconscious activity of the mind consists in imposing forms upon content' and 'these forms are fundamentally the same for all minds [. . .].'[5] This imposition of order is said to be a distinguishing mark in the movement from nature to culture. Ernst Gombrich goes a step further, widening the argument to include all living organisms. They are pre-programmed to take appropriate action rather than learning only from experience: 'These actions pre-suppose what in higher animals and in man has come to be known as a "cognitive map", a system of co-ordinates on which meaningful objects can be plotted.'[6] It is not necessary here to resolve the question of how universal this tendency might be. At the cultural level, however, maps are invariably used to impose meaning on the world. Otherwise bewilderingly complex and unwieldy masses of phenomena are carved up into manageable portions through the imposition of various grids like those used to map a territory, grids that create the reality they often appear merely to represent.

Language is among the most powerful of these mappings. Different languages impose very different grids and meanings, chopping up the flow of experience in different ways.[7] Meaning comes from the map imposed on the territory rather than from the territory itself, although a host of extra-linguistic forces may also be involved. The linguistic grid becomes sedimented into experience, taken to be an objective map of a prior reality rather than an arbitrary imposition. Each language contains in its interstices an implicit philosophy, a metaphysics or a world-view. These can differ widely. The Hopi culture of North America, to give one classic example, has no notion of time as a flowing continuum, a concept fundamental to our underlying view of reality yet specific to particular cultural frameworks.[8] Western notions of time and space have themselves undergone change in the face of successive technological and cultural dynamics associated variously with what have been termed 'modern' and 'postmodern' experiences. As Edward Sapir puts it: 'No two languages are ever sufficiently similar to be considered as representing the same social reality. The worlds in which different societies live are distinct worlds, not merely the same world with different labels attached.'[9]

The various continua of experience may be divided up in many different ways. Different cultures may separate out different individual colours from the continuum of the spectrum or different cardinal points of direction. Arbitrary divisions are also imposed as we negotiate our way through different stages of life. The physiological process of growth from birth to maturity is largely continuous. Yet cultures impose a structure upon this movement that rationalizes development through a series of discrete phases such as childhood, adolescence and adulthood. Where the identification of age groups is important to the organization or understanding of a society the imposition of such structures may help to reduce confusion. Rather than merging into one another, distinct stages of life are institutionalized. These may be linked to physiological developments but more arbitrary cultural factors tend to predominate.[10] The boundaries between different stages often take on great symbolic significance. Rites of passage are employed to negotiate these transitional spaces and tackle the ambiguity or anxiety they might otherwise cause. Instead of being allowed gradually and imperceptibly to move from the status of adolescence to adulthood, for example, members of a society may at a certain clearly defined point undergo ritual initiation from the one to the other. This will often involve a moment of separation from the social mainstream followed by a period of transition before rituals of incorporation in which individuals rejoin society in their new roles. Various strategies are often available to deal with ambiguous cases that might otherwise threaten to undermine any system of classification. Accusations of witchcraft and heresy are often connected with matters of ambiguity or nonconformity. That which might otherwise be subversive can be projected onto another plane of reality, leaving the conceptions of the ordinary world relatively untouched.[11]

All classificatory grids are arbitrary. They have no necessary or absolute status. It does not matter what kind of grid is used on the map. Any system of lines and points of reference can be imposed to provide orientation, although different mappings may serve very different interests. This is not to say that they are in any way trivial or easily shuffled off. This is an important distinction. *In their own context,* these structures are far from arbitrary and may, indeed, be experienced as more or less absolute or necessary. For those who inhabit particular mappings they are likely to be viewed simply as reality. To those born into them they tend to appear natural. Cultural grids are not generally seen for what they are. Their mappings are assumed to exist on the territory itself, as 'the way things are' in a taken-for-granted common sense, although this is not to say that they are never contested. Particular constructions of the real or the objective result from the failure to recognize the provisional status of such maps. Pierre Bourdieu criticizes Lévi-Strauss, among others, for converting these structures into abstract geometrical quantities when they remain the outcome of active cultural-historical processes. It is for this reason that he questions the suitability of the mapping metaphor I have used:

> It is significant that 'culture' is sometimes described as a map; it is the analogy that occurs to an outsider who has to find his way around in a foreign landscape and who compensates for his lack of practical mastery, the prerogative of the native, by the use of a model of all possible routes. The gulf between this potential, abstract space [. . .] and the practical space of journeys actually made, or rather of journeys actually being made, can be seen from the difficulty we have in recognizing familiar routes on a map or town plan until we are able to bring together the axes on the top of the field of potentialities and the 'system of axes linked unalterably to our bodies, and carried about with us wherever we go', as Poincaré puts it, which structures practical space into right and left, up and down, in front and behind.[12]

Ethnologists, Bourdieu suggests, are often dependent on a limited number of sources and badly positioned to tell the difference between official and unofficial accounts, the mapped route or the beaten track. The construction of abstract networks, such as the kinship structures drawn up by Lévi-Strauss, may obscure the real meanings such relationships have for those involved. Resistances, alternatives and ulterior motives may be left out of the picture: 'The logical relationships constructed by the anthropologists are opposed to "practical" relationships – practical because continually practised, kept up, and cultivated – in the same way as the geometrical space of a map, an imaginary representation of all theoretically possible roads and routes, is opposed to the network of beaten tracks, of paths made ever more practicable by constant use.'[13] Rather than detailing frozen mathematical structures we should examine the ways these orders are produced. In this way it is possible to answer those critics of structuralism[14] who accuse it of denying or ignoring any principles of historical change. Cultural mappings are only ever *naturalized*, never natural. This also applies to the structure that generates our own contemporary Western concept of reality as conventionally distinguished from the image, representation or map. The distinction made here is just another way of chopping up a continuum, no different in principle from those described above. Its development, as it happens, may also owe a debt to cartographic techniques.

The rediscovery of Ptolemy's *Geographia* in the early fifteenth century played a central role in the establishment of Renaissance principles of linear perspective, a forerunner of contemporary constructions of pictorial realism. The ostensible aim of the artists and scientists involved in this enterprise was to achieve objective representation of the world. This was contrasted with earlier medieval perspectives based on different conventions of representation. Medieval maps tended toward what we might now describe as an impressionistic depiction of features such as settlements and hills rather than claiming to be mathematically objective.[15] Samuel Edgerton demonstrates the difference between medieval and Renaissance representations by comparing two map/pictures of Florence. One, made before the development of linear perspective, depicts the city as a crowded jumble of buildings in no apparent order. The second, the *Map with a Chain* (unknown artist, *c.* 1480), benefiting from the new techniques, appears to be a more 'rationally' ordered geometric design, although it does not conform strictly to the rules of linear perspective. But, as we have already seen, such developments were not a simple matter of increasing objectivity or accuracy. The two maps conform to different codes of representation rather than one being inherently superior or inferior to the other. Two different ways of experiencing the city are recorded, each with its own merits. While establishing a particular rhetoric of objectivity, the *Map with a Chain* locates the viewer in an abstract space alienated from the city. Its predecessor might give a better feel of the plastic or sensual aspect of Florence, albeit one of limited use for navigation.[16]

Artists who adopted linear perspective techniques used a grid system to lay out the design, a principle similar to Ptolemy's cartographic system for mapping the surface of the globe through a network of horizontal and vertical lines. The difference between a painting, the view of a Renaissance city or a map of the world was one primarily of scale.[17] In each case the grid system permitted the heterogeneity of the world to be reduced to a geometrical uniformity. The rediscovery of Ptolemaic maps played an important part in an apparent rationalization of the world. Earlier maps such as portolan charts may have been valuable for seamen but their scope was limited. Now it seemed that the entire globe could be comprehended within a single framework, at a time when much of the world remained unexplored. It was not long before the abstract grid was used for the marking of geographical boundaries.[18]

Changes in the nature of mapping in this period are associated with a complex series of historical movements. According to a conventional and inevitably over-simplified historical mapping there was a shift from the medieval world to that known as the Renaissance and later the Enlightenment, terms that are themselves imbued with their own teleological assumptions of objectivity and scientific impartiality. The historical moment was also one of transition from a feudal to what would become a predominantly capitalist form of economy and society in the West. The medieval world is usually seen as a closed one. While local maps or views might have had a more plastic appearance, the map of the universe was rigidly fixed. The earth was enclosed at the centre of a series of crystalline spheres containing the planets and at the outer rim the fixed stars. Beyond the stars was the ultimate level of Heaven while beneath the earth were the rings of Purgatory and Hell, as depicted in Dante's *Divine*

Comedy. The terrestrial world was held by dominant theological constructions to be a microcosm to which an equivalent social mapping was applied. For many people these cartographies provided a firm foundation for existence, although it would be wrong to assume that they were passively accepted in all cases. In the West the known world was a small one, centred on the Mediterranean and bounded by uncharted seas.

New and different forms of mapping were developed to meet various needs, not least those associated with the spread of a money and commodity based economy. Both time and space underwent a process of rationalization. Detailed land maps were essential when land became an alienable product, to be bought and sold, along with all others. The town gradually began to overtake the dominance of the countryside, a process marked by the appearance of map-like views such as the picture of Florence described by Edgerton or Barbari's view of Venice. A new, abstract space was produced. With the work of Newton this same rationalizing process was applied to the structure of the entire universe. No longer the domain solely of religious dogma, it too was embraced by an abstract mathematical grid.

It would be mistaken too quickly to dismiss the influence of religious or any other ideology, as we saw in the preceding chapter. The development of linear perspective might appear to represent a victory of objective reality over medieval mysticism. Yet the rational and the mystical were aspects of the same cultural complex. At the time of its development linear perspective was in fact viewed as a means of restoring the flagging moral authority of the church in an increasingly secular world. A world of straight lines and mathematical homogeneity was seen as reflecting the perfection of God's creation. Linear perspective is just another cultural convention. It might have its uses, but it is not the only way to view the world.[19]

Geometrical grids were also adopted by a number of twentieth-century avant-garde modernist artists, including Piet Mondrian and Agnes Martin. Although a repeated stereotype, the grid was again mobilized in the name of authenticity, an 'originary' purity marking an apparent break from past traditions, as Rosalind Krauss suggests. The image of the grid was proposed as the manifestation of 'an indisputable zero-ground beyond which there is no further model, or referent, or text.'[20] But this is a fiction:

> The canvas surface and the grid that scores it do not fuse into that absolute unity necessary to the notion of an origin. For the grid *follows* the canvas surface, doubles it. It is a representation of the surface, mapped, it is true, onto the same surface it represents, but even so, the grid remains a figure, picturing various aspects of the 'originary' object: through its mesh it creates an image of the woven infrastructure of the canvas; through its network of coordinates it organizes a metaphor for the plane geometry of the field; through its repetition it configures the spread of lateral continuity. The grid thus does not reveal the surface, laying it bare at last; rather it veils it through a repetition.[21]

At the same time the grid also comes *before* the empirical surface of the painting, 'preventing even that literal surface from being anything like an origin.' The map both follows and precedes the territory. Behind the grid lie a whole range of prior texts according to which it was organized.

> The grid summarizes all these texts: the gridded overlays on cartoons, for example, used for the mechanical transfer from drawing to fresco; or the perspective lattice meant to contain the perceptual transfer from three dimensions to two; or the matrix on which to chart harmonic relationships, like proportion; or the millions of acts of enframing by which the picture was reaffirmed as a regular quadrilateral. All these are the texts within which the 'original' ground plane of a Mondrian, for example, repeats – and, by repeating, represents. Thus the very ground that the grid is thought to reveal is already riven from within by a process of repetition and representation; it is always already divided and multiple.[22]

Neither linear perspective nor the modernist grid have any unique claim to authenticity or objective representation of the world. If realism is taken to mean that which best grasps a given reality it cannot be restricted to a particular set of formal techniques. Neither can it make any transcendent claims. As

Bertolt Brecht argued in his polemic against Georg Lukács: 'We must not derive realism as such from particular existing works, but we shall use every means, old and new, tried and untried, derived from art and derived from other sources, to render reality to men [*sic*] in a form they can master.'[23] In the Renaissance world linear perspective may have been the most appropriate expression of the dominant reality, just as the nineteenth-century realist novel may have seemed to Lukács the best way to convey the texture of his times. But elsewhere different forms might be required. No single form of representation has any absolute claim to objectivity.

Ambiguous cases which challenge the terms of existing mappings may in some cases be welcomed and embraced. Often the response is a strong reaffirmation of existing boundaries or the establishment of new ones. This is particularly true of dominant Western constructions of sexuality, a map that impinges intimately on our daily lives. Neither physiologically nor culturally do all humans fall simply into the two sexual poles of male and female. A wide variety of intersexual possibilities exist. Yet a reductive binary mapping is imposed. We are capable of erotic responses across the continuum of the physical body. Patriarchal definitions of sexuality tend, however, to enforce a body-mapping that limits the erotic to the experience of the genitals and a number of other 'erogenous zones'. Rather than absolute and given entities, masculine and feminine are cultural constructs, reductive fictions to which reality is made to conform. The rigidity with which this mapping is imposed is at least in part a response to the lack of any ultimate grounding, particularly where it provides the basis for a system of domination. Here again we see a level on which these ultimately provisional structures are far from arbitrary in their effects. They can serve very specific interests.

The fact that dominant constructions of sexuality are essentially fictional does not make them any the less real in their impact. They provide a powerful framework through which is defined much of the social reality we inhabit, rather than being some kind of afterthought tagged on to a more fundamental ground. It is at precisely such levels that sexuality is real to us and establishes the territory on which we live. The material basis that helped to support this social construction in some cultures may have subsided to a certain extent. Changes in working patterns and associated work-cultures resulting from the relative decline of traditionally 'masculine' industries in some areas might be expected to create conditions in which a more fluid construction of male sexuality could develop. But there is little evidence that this has happened.[24] Gender structures might be defended all the more keenly in response to the dissolution of other social frameworks. A recurring image in dominant Western male sexuality is the defensive construction of a fortress-like masculinity designed to keep out what is perceived as the threat of women: 'She is the landscape that we admire, yet tread warily across. She is to be feared and controlled. She is the vast mother with a cave between her legs. She embraces us when we are infants and ever since threatens to swamp us and swallow us.'[25] The outcome of such a fear may be a violent repression and objectification of women in the reassertion of masculine control.

It may seem paradoxical to describe widespread male violence against women in defensive terms, but in patriarchal societies masculinity has a position of dominance and advantage to protect, a defence that is likely to be all the more oppressive the more shaky its ultimate ground. There are other circumstances also in which the imposition of rigid cartographies can be a way of securing or maintaining the domination of the powerful. This may be said of almost all languages and maps of meaning. They can never entirely be divorced from the networks of power they both facilitate and support. More literally, geographical territories may be carved up to secure control, although the imposition of clearly marked and precise *national* borders is a relatively recent phenomenon. Many earlier empires gained effective control over large dominions through a more amorphous structure in which borders were 'porous and indistinct, and sovereignties faded imperceptibly into one another'.[26] The use of rigid grids has become a classic tactic of modern colonialism. A graphic illustration is provided by colonial strategy in the Arabian Gulf. Before colonial rule the area was not governed according to the particular concepts of sovereignty over land that became dominant in the West. Tribes migrated across wide areas. There was a flow of both trade and people and a great deal of flexibility. Much of this was ended under colonial regimes. The area was divided into a number of distinct states whose existence was arbitrary in any terms other than those of domination and the exploitation of resources by colonial powers and their local allies.

> The notion of the state in the modern sense was introduced arbitrarily in order to further British interests: a new system of political rule was imposed on the pre-existing social system, with tribal chiefs now being forced to accept legal responsibility for the members of all tribes that owed them allegiance. This accounts for the chessboard division of the emirates and the various neutral zones between them, and the difficulty of defining borders up to the present day. Border disputes sharpened with the initial oil discoveries after the First World War, and were critical in frustrating the case of unity.[27]

Divide-and-rule tactics ensured that no single state would become large enough to challenge the colonial power. Territories of great strategic value (initially because of their location on trade routes to India, later because of the discovery of oil) were cut off from larger units that would be less easy to control. The legacy is felt today in continuing territorial disputes. The most recent manifestation was the Iraqi invasion of Kuwait in August 1990. However distasteful its regime at the time, Iraq had for many years laid claim to a well-endowed neighbour maintained principally as a 'rotten borough' controlled by Western powers.

Maps of the Middle East have been used by all sides in the pursuit of claims to territory. Others were published during the crisis in the Gulf to provide orientation for the general reader. Rand McNally hurried out an updated chart of the region under the title 'The Desert Storm War Zone Map', while Hammond countered with its 'Crisis . . . Middle East'. Each marked with dotted lines those national boundaries accepted by the Americans to be undefined or disputable. These did not, of course, include the line between Iraq and Kuwait, although anyone examining the maps might be puzzled by some of the demarcations, such as the 'neutral zone' between Iraq and Saudi Arabia. Maps were used widely in media accounts of the war, often giving a rhetorical impression of understanding where detailed knowledge was lacking.[28] Arrows marking movements or threats could be used on such maps, variously, to show Iraqi strength or aggressive intent, to justify the use of force by the United States and its supporters or to show the speed and efficiency of the US assault for purposes of future deterrence. Iraqi cartographers were quick to depict Kuwait as the country's nineteenth province, just as Israel had produced maps to support its appropriation of the land occupied during the Six-Day War in 1967. One Israeli map showed its tiny land surrounded by a huge black mass of Arab states. Although technically accurate, the map was a work of propaganda designed to counter accusations that Israel was the aggressor. A comparison on the basis of land area said nothing about the advanced military technology possessed by Israel or its alliance off the map with the United States.[29] It also gave a misleading impression of Arab homogeneity. From the other side, a collection of interviews and speeches by President Assad of Syria distributed during the peace talks of April 1992 had a map of the region on the cover in which Palestine was reinstated in place of Israel two years before elements of the reality were established with the creation of autonomous zones in the occupied territories.[30] In areas such as the Middle East, where borders are particularly fragile, the questioning of one may threaten all, as Keiller suggested in *Lake Wobegone Days.*

Ultimately the boundaries of all nation-states are arbitrary, however passionately they might come to be attacked or defended. A comparison between physical and political maps of the world shows relatively few borders determined principally by geographical terrain. Even where they are the very concept of nationhood remains a recent one and there is no guarantee of their permanence. Boundaries may be all the more vigorously defended when their geographical or social basis is particularly weak. Israel is a classic case, carved out of Palestinian land on the basis of a biblical mapping taken from a period of some sixty years two thousand years ago.[31] A parallel could be drawn between this arbitrary mapping of the physical territory and the kind of cultural boundaries established by the eating prohibitions laid down for the Jewish people in the book of Leviticus. What is forbidden is that which transgresses the classificatory grid laid down in the Bible.[32] Rather than being derived from any particular characteristics of the foodstuffs concerned, the prohibitions may have served simply to differentiate the Jewish people, to make them stand out culturally from hostile neighbours.

The example of the Middle East offers a salutary reminder to those who champion the supposed arrival of a postmodern 'New World Order' in which national boundaries are said to be blurred by an increasingly global economics and culture. Instead we find a dialectical movement between the dissolution and reassertion of boundaries, the erasure and the reinscription of lines on the map. The role of the United Nations in the Gulf conflict in 1990–91 led some to celebrate the apparent fulfilment of the dream that it could become a genuinely independent global police force, freed from the constraints of Cold War power politics.[33] The reality was rather different. The claim that the UN had only been prevented from realizing its ideals in the past because of a Soviet veto in the Security Council is misleading. The United States had repeatedly used its veto (or abstentions) to block resolutions seeking to uphold international law, from Central America to the Middle East and elsewhere.[34] If the war in the Gulf was heralded as marking the arrival of a new era, the American invasion of Panama in December 1989 demonstrated how little had changed in the practices and assumptions of leading American policymakers, even as the dismantling of Stalinism in Eastern Europe was at its height.

The 'New World Order' is if anything one of increased US hegemony, partly as a result of the inability of Russia or the Confederation of Independent States which replaced the former Soviet Union to maintain more than a token influence beyond their own fragile borders. In this respect it marks the achievement of a long-standing goal of American foreign policy: to keep as much of the world as possible open as a supplier of raw materials, a source of cheap labour and a market for manufactured goods; a world mapped out as far as possible into a single economic zone in which many states are reduced to playing subordinate roles rather than establishing their own manufacturing or trading blocs. Others have also benefited. The American postwar effort to ensure that Japan and Germany remained wedded to the international capitalist order helped them eventually to overtake the United States in terms of economic strength. Other challenges come from the economies of Hong Kong, Taiwan, South Korea and Singapore. The greatest beneficiary may have been international capital itself, a global form that is not rooted in any one particular national soil. But the dominance of multinational capital does not entail the demise of the nation-state. Multinational corporations need political frameworks of authority and control for the furtherance of their own ends. America remains militarily dominant and thus best able to secure a global environment geared towards their needs. As a result it might be expected to shift confrontation into the arena in which it is strong, undermining rather than championing the diplomatic initiatives of the United Nations, as appeared to be the case in the Gulf. Similar doubts afflict celebrations of globalization at the cultural level. New boundary-crossing media tend to increase the cultural hegemony of the Western image-machine rather than to allow the expression of any widespread diversity, although they can in some cases be put to different use. Some borders remain firmly in place, most notably the unbridged North–South divide between most of the rich and poor nations of the world. Initiatives such as the North American Free Trade Agreement may appear to blur key national boundaries such as that between the United States and Mexico. But they also underline the importance of the maintenance of boundaries for those who want to take advantage of cheap labour or weak environmental regulation on the other side of the border.

A dialectical movement between the dissolution and reimposition of boundaries is seen nowhere more clearly than on the recent map of the Soviet Union. For decades the outline of the USSR was blocked in boldly, a huge frozen mass stretching across northern Europe and Asia. The area included within its orbit was increased after 1945 with the addition of client states in Eastern Europe. The whole bloc became reified, a brittle structure that with only occasional and limited exceptions took on the appearance and character of a solid reality. Suddenly, in the late 1980s and early 1990s, it began to crumble. The map became alive, mobile and unstable. Lines that had appeared sharp and inviolable were blurred. The same could be said of the conceptual mappings according to which the territory was understood. Many on the right found a supposed vindication of capitalism and a confirmation of their preconceptions. A new zone on the map was opened up to the penetration of Western capital. The left was disoriented, torn between celebrations of democratic advance and the fear that certain social gains in Eastern Europe were likely to be lost in a mad rush into the 'free market'.

So great was the redrawing of the map that all boundaries might have appeared questionable. But others were asserted. Rejuvenated nationalisms broke out across the map of the former Soviet bloc. It did not become a zone of fluidity. The fifteen republics of the Soviet Union appeared on the map in their own right, new and unfamiliar presences to most in the West. Other ethnic groups also asserted their rights, at an ever-diminishing scale that again emphasized the arbitrariness of national borders. The last days of the Soviet Union were marked by abrupt swings between fragmentation and the imposition of renewed central control. Examples of the latter include the failed hardline coup of August 1991, Boris Yeltsin's assault on the White House parliament in October 1993 and the subsequent electoral success of the neo-fascist Vladimir Zhirinovsky. Zhirinovsky has demonstrated his desire to extend the borders of Russia up to and beyond those of the USSR, taking up a pen in a television interview to draw new expansionist lines on the pages of an atlas.[35]

Yugoslavia entered a state of civil war and collapse. Rather than being blurred, the lines on the Yugoslav map were reinscribed in blood. Previously unmarked borders between the six Yugoslav republics reappeared under the pressure of militant Serb nationalism and resultant declarations of independence from Slovenia, Croatia and Bosnia-Herzegovina. The separate republics also began to appear on international maps. Each rival group attempted to redraw the map according to its own designs. The Serb nationalist map laid claim to any territory occupied by Serbs. The autonomous provinces of Kosovo and Vojvodina, which had been given effectively the same status as the republics under a new constitution in 1974, were swallowed up by the Serbian state. Having failed to prevent the break-up of the centralized Serb-dominated federation, the Serbian leadership under Slobodan Milosevic sought to impose its nationalist map of a Greater Serbia onto reality.[36] An 'Autonomous Province of Krajina' was written unilaterally onto the map in Serb-dominated eastern Croatia, followed by further Serb autonomous regions in Bosnia-Herzegovina. Croat nationalists issued their own map of a Greater Croatia, which went on sale in Zagreb, and responded to Serb initiatives by inscribing their own zones onto the map of Bosnia-Herzegovina. Closely intermingled populations of different origins in Bosnia were faced with a violent remapping onto the territory along ethnic lines, a process that became known as 'ethnic cleansing'.

The map itself became a key battleground in the debate over how Bosnia-Herzegovina might be dismembered. Rival Serb, Croat and Muslim delegations haggled and fought over a series of maps proposed by the United Nations and other groups of intermediaries. A confusing array of maps accompanied the reporting of the process in the Western media. Newly arrived colour printing technology provided some newspapers with the means to show clearly the complexity of both the ethnic composition of the former Yugoslavia and the various proposals and counter-proposals for the redrawing of the map. The rhetorical impact was rather different. Instead of helping to create understanding, such maps tended to reassert the complexity and supposed irrationality of events in the Balkans, as was suggested by some of the images used to describe them – resemblance to a Jackson Pollack painting or an ink-blot test, for example.[37] Economic and class strains underlying the conflict went largely unmapped and unreported, while little attention was paid either to the extent that the differences between the warring factions were the stuff of culturally imposed nationalist mappings rather than supposedly timeless 'ethnic' essences.

Where one bloc is fragmenting, others may take its place. Thus the former Soviet republics may be drawn together into new relationships either with one another in the Confederation of Independent States or with a European Union that is itself torn between the dynamics of a deeper and closer membership or a wider-ranging one. The newly separated republics of Yugoslavia are likely to come under the influence of new, or older, gravitational forces: Croatia and Slovenia pulled towards Germany and Austria, Serbia towards Russia and Greece, and the Muslims towards those who have expressed solidarity with them from the wider Muslim community.[38] The new map of Europe is in flux, but the movement is complex and dialectical. Deeper unities are being forged by the European Union harmonization programmes launched in 1992. In response, reassertions of cultural identity and national borders come from the neo-fascist and more respectable elements of the right; from those on the left who see the increasing powers of Brussels and Strasbourg as diminishing the prospect of the achievement of socialism in their own countries; and from security services advocating the introduction of stricter controls over populations allowed to move more freely across state frontiers. Midnight at the start of the New Year of 1993 saw, simultaneously, the dismantling of borders across the European Union and the construction of a new one between the Czech Republic and Slovakia.

A regional emphasis seems to have emerged within the wider framework of the EU. New regional structures have been created in recent years in France, Italy and Belgium.[39] In Britain, an increased degree of autonomy for Scotland at least is back on the political agenda: Scottish and Welsh nationalists talk less of outright autonomy than of independence within a wider EU framework. The European Union itself has forged direct links with regions which, along with other cooperative ventures, effectively bypass national governmental structures. A map compiled according to these principles depicted the continent as a colourful mosaic:

> The map of Europe – the map of the real Europe, as opposed to the conventional lines and colours – is changing utterly. In 1914, it was a simple diagram of a few large nation-states and empires. In 1992, it is coming to resemble a patchwork quilt, and a very badly sewn one too. As maps go, all Europe is coming to resemble the atlas of old Germany before 1871, when it was a multicoloured mess of tiny princedoms, Hanseatic city-states, kingdoms and grand duchies.[40]

4 Mapping the Void

So Geographers, in Afric-maps,
With savage-pictures fill their gaps;
And o'er uninhabitable downs
Place elephants for want of towns.

Jonathan Swift[1]

The postmodern experience has been described as a blurring or fragmentation of the outlines on the map. Yet many existing contours have been blocked in all the more boldly. Religious, cultural and political fundamentalisms continue to thrive. Both modernity and postmodernity have, in various accounts, been depicted as a sweeping away of existing cartographies. In Marx and Engels' classic phrase, adopted by Marshall Berman as the title of his study of modernity, 'all that is solid melts into air'[2] in the face of the ongoing rush of change under the relentlessly revolutionizing logic of capitalist forces and relations of production. For Marx this process entails the reduction of everything to the particular narrow rationality of capitalism. Professionals are reduced to the status of wage-labourers; familial relations and national differences are obliterated; law, morality and religion alike are stripped of their facades. There is also a countermovement, however, equally important to the maintenance and reproduction of the regime: a reassertion of structures where others have been dissolved, a redrawing of the lines on the map rather than a submission to the void.

Capitalist and various modern or postmodern dynamics may have swept away a multitude of earlier frameworks, such as the mutual obligations and clearly defined status relationships of feudalism. But the formations that result need their own limiting structures. Hence the sometimes forcible imposition of new or altered frameworks, particularly around constructions of sexuality, race, nation, 'the family' and the like, a process of 'deterritorialization', as Gilles Deleuze and Felix Guattari put it, in which the surface of the map is rendered blurred and slippery, followed by one of 'reterritorialization', when new grid references are inscribed.[3]

Two opposite responses might result from a perceived crisis of existing boundaries.[4] A last-ditch effort is often made to defend or to strengthen the lines on the map. Where geopolitical or other change is rapid, and particularly where it is seen as a movement towards dissolution, it is likely to be met by a strict reinforcement of cultural frameworks. The destabilizing effects of change may be countered, whether for the furtherance of narrowly dominant or wider interests. An obvious example is found in assertions of religious fundamentalism, both Christian and Islamic, or its political equivalent trumpeting 'Victorian values' or a movement 'back to basics'. Fundamentalism presents itself as a reversion to basic values in the face of what is seen as an increasing decadence. It can also be used by others to assert their own position against forces of change. The United States in particular tends to create a simplistic and reductive image of Islam that serves a number of domestic purposes.[5] Islamic fundamentalism is presented today as one of a number of new threats to replace that of the former Soviet Union and as a justification for continued military spending.

Alternatively, the process of deterritorialization may be encouraged. Deleuze and Guattari advocate that the disintegrative movement of capitalism be pushed to its logical limits. Rather than accepting the bounds of superficially imposed structures they propose a letting loose of flows of desire. Instead of a psychoanalytic referral of all psychic problems to a triangular Oedipal mapping, Deleuze and Guattari offer 'schizonalysis', an encouragement of tendencies towards disintegration that seems to offer little realistic prospect of a model for tolerable existence.[6] The problem with this position is found at two different levels in accounts of the postmodern. In Deleuze and Guattari's *Anti-Oedipus* (1972), the difficulty occurs at the level of the strategy prescribed for the future rather than the analysis of the existing situation. Others seem to misread what they describe as the postmodern. Thus for Terry Eagleton, it is the 'depthless, styleless, dehistoricized, decathected surfaces of postmodern culture'[7] that mark its wholly unambiguous implication in capitalist consumer society. In arguing for the importance of resistances to such implication Eagleton allows a slippage that implies that it is the postmodern era itself that exhibits such flattened surfaces. But this is a reductive mapping of far more ambiguous and contradictory phenomena. The point is not that existing cartographies are being strictly maintained or dissolved, or that they should be. We have to recognize both their provisionality and the fact that we need some kind of mapping within which to orient our lives.

Accounts of the postmodern that describe it in terms of unmappable surfaces fail to allow for the redrawing of lines on the map and to recognize such a dialectic more generally in the creation of cultural meanings. It may be no coincidence that the most sophisticated early maps were drawn in relatively featureless universes like those inhabited by Eskimos and the inhabitants of small islands in the Pacific. The typical cultural response to a lack of existing features is to impose mappings that create meanings and so make the territory negotiable, both physically and conceptually, rather than to submit to an undifferentiated existence. For Jean Baudrillard, the giant buttes of Monument Valley, instantly recognizable as the setting for the westerns of John Ford, can be likened to 'blocks of language' carved out of the landscape. According to Baudrillard these, along with all other meanings, are being eroded in a process of cultural desertification. We could use Baudrillard's image differently, however. The American deserts, he suggests, 'denote the emptiness, the radical nudity that is the background to every human institution. At the same time, they designate human institutions as a metaphor of that emptiness and the work of man as the continuity of the desert, culture as a mirage and as the perpetuity of the simulacrum.'[8] It is true that human cultures and institutions stand upon little firmer ground than the void evoked by Baudrillard, but they are not so easily worn away. Where they are eroded they are likely to take on different forms or be replaced by others.

An illustration of this process is found in the account of the reclaimed English fenland landscape in Graham Swift's novel *Waterland* (1984). Left to themselves the fens would be undifferentiated and featureless, a world subjected to the levelling effects of the water, 'a liquid form of Nothing'.[9] The map would be empty. The world Swift describes appears to be subject to the second law of thermodynamics, a steady decline into entropy. This view is akin to the early Christian concept of precarious life on a dry land that would remain beneath the seas but for a special arrangement for humanity made by God. Never fully reclaimed, the fens are only ever *being* reclaimed. Without the constant work of pumps, dykes, embankments and dredgers, the landscape would revert to a state of watery equilibrium. No truly solid ground is available, but this does not mean that diligent work cannot create something firm enough upon which to stand.

Cultural mappings establish a similar kind of reality. The fens were among the earlier parts of Britain to be mapped, because of the demands of drainage: a creation of meaning in the landscape just like that found in other cartographic works. The ancient Egyptians have been credited with inventing geometry as a response to the demand for repeated property surveys and maps for taxation purposes when the periodic flooding of the Nile erased existing boundary markers. The watery Venice mapped by Jacopo de Barbari offers another image of a cultural landscape defended against the threat of erosion. Such imagery could be applied directly to the accounts of male sexuality given in the previous chapter, especially the notion of woman as a liquid presence that threatens to dissolve its strength. Not the immediate reflection of immanent structures 'out there', as givens, in the world, or the expression of any essential subjectivities, such cultural frameworks are more or less arbitrary maps imposed upon the territory, conceptual frameworks that enable us to map our position in what might otherwise be a disturbing void. As Linda Hutcheon puts it, such structures 'are human constructs in history. This does not make them any the less necessary or desirable. It does, however [. . .] condition their "truth" value [. . .]. The point is not entirely that the world is meaningless, but that any meaning that exists is of our own creation.'[10] The reality we inhabit is of our own making.

The strategy advocated by Deleuze and Guattari, to submit to a fluid, unmapped existence, does not appear to be open to the inhabitants of Swift's fens. That would mean facing up to a reality that is 'uneventfulness, vacancy, flatness. Reality is that nothing happens.'[11] An existence of such unrelenting monotony may be unbearable, 'enough of itself, some might say, to drive a man to unquiet and sleep-defeating thoughts'; 'Melancholia and self-murder are not unknown in the Fens. Heavy drinking, madness and sudden acts of violence are not uncommon.'[12] The world of the schizophrenic, like that of the open fens, may offer certain new horizons or an escape from overly restrictive frameworks. The same may be said of the possibilities opened up by less (or differently) structured cultural mappings. There may be moments of existential insight into the void that lies behind cultural cartographies, but the price to be paid is high. Moments in which we see through the structures that envelop us may be as much terrifying and disabling as enlightening. As Paul Carter puts it:

> We begin to feel both the horror of the plains [or fens] and their attraction, a tension that, perhaps, only the map-making mind can bring under control. For if the horizontal is [. . .] open in all directions, the place where all futures are possible, then it is also the focus of annihilation, the inescapable point of return, where all variety is reduced to a level. In this latter aspect, it signifies the waste of vital energies, the dissipation of promise, the oblivion of an imprisonment without walls. This is the double aspect of the plain: that it releases, but releases into nothingness. Directionless and equal, it inhibits motion. It resists exploration.[13]

An imperative to create familiar order in strange new landscapes was felt by many explorers of the New World. Matthew Flinders, whose account *A Voyage to Terra Australis* was published in 1814, came from the fens. Onto the Australian landscape he mapped the names and layout of a number of Lincolnshire towns and villages; a mapping imposed on one otherwise undifferentiated landscape was used to bring another under apparent control.[14] Such strategies were essential not so much to the discovery as to the *invention* of the Australian landscape. The very existence of Australia had long been either proclaimed or denied in the West at the level of a theory unadulterated by actual exploration. The existence of a much larger *Terra Australis* was asserted long before the travels of Cook or Flinders, on purely formal grounds. A map of 1493 by Macrobius depicts a round world the upper half of which – comprising Europe, Africa and Asia – is balanced below by a large '*Teperata-Antipodum*' or '*Nobis incognita*'. The 1587 map of the world by Ortelius is in general much more modern and recognizable to us, but the huge hypothetical southern landmass remains, a '*Terra Australis Nondum Cognita*'. Abstract grounds were also used for the denial of its existence by flat-earthers or those who imposed a biblical mapping that could not allow the possibility at least of an inhabited land beneath what was widely accepted to be an impassably torrid equatorial zone: it would be incompatible with the belief that the earth had been populated through the dispersal of the sons of Noah.

Explorers who eventually found the real Australia imposed their own mappings on the territory. The early maps and journals were filled with misnomers: 'meadows' and 'mountains' that owed little in appearance to what usually went by the names, but that provided forms of spatial punctuation.[15] The English language lacked the words in which to characterize an alien landscape in its own terms, but the problem was more than simply one of naming pre-existing features. They were to be *constructed* according to a rhetorical process of ordering and claiming the territory. Important signposts on the map such as mountains, hills and rivers were scarce in parts of the Australian interior. Where they did not exist on the ground they might be invented, translating a hostile landscape into a legible text. The less there was for the explorer to see, Carter suggests, the greater might be the need not just to write *about* but to *write* the terrain itself. To designate a feature such as a 'mount' might precisely be to express the absence of such a feature as usually understood. In some cases the process went a stage further. One surveyor working in the sand dunes of the Victorian Mallee is reported to have constructed his own hills to make up for the shortcomings of nature.[16] Arbitrary mappings stuck: 'They were nothing more than names and outlines on maps. They bore no relation to reality, but without them travelling was impossible. Whether or not they deceived with their promises of water and anchorage, they did not deceive in harbouring "resting places for the imagination".'[17] The meaning of such landscapes is neither something given objectively by landforms themselves nor a matter of free individual interpretation. It is an essentially collective, cultural meaning, found at both macroscopic and microscopic social levels, on the map of the whole territory and in the multiplication of symbolic boundaries at the domestic scale.

A further semblance of control came with the imposition of a universal taxonomical grid devised by Carl Linnaeus and applied to the Australian territory by the botanist Joseph Banks who accompanied Captain Cook. The Linnaean system of the eighteenth century offered a ready-made framework into which new discoveries could be slotted, regardless of their novelty.

The basis of the grid was arbitrary, a fact that Linnaeus himself acknowledged.[18] He chose to concentrate on the flower parts of plants, classifying them accordingly and ignoring other similarities or differences. Four variables were selected as the basis for categorization. A potentially overwhelming plenitude was thus reduced to the confines of a scheme into which every variety could be fitted. For Michel Foucault the grid imposed on the botanical world was akin to those developed at the same time to achieve a similar kind of domination in the prison, the military barracks, the hospital and the workplace.[19] In each case the potentially unruly was kept in line.

Maps tend, if not always to fill in, at least to enclose what might otherwise remain blank spaces, such as those so fearful to the inhabitants of the imaginary land described by Peter Carey. Many instances have been documented of cartographers using their imaginations to fill what would otherwise be blanks on the map, as in Swift's quatrain. Blank spaces are there to be filled, however unlikely the results. In the history of the early mapping of America we find an overlapping series of fantasies and inventions that served to fill the unexplored spaces of the western and central areas. Wherever factual detail was absent the imagination could be called into play by cartographers, whether the intent was devious or innocent. Unexplored lands were filled with pictures of animals, real or fantastic, or with invented rivers and mountains. Uncharted seas were inhabited by mermaids, sea monsters or islands of the imagination.[20]

Onto such creations a variety of sentiments could be inscribed. Rumours quickly became established as facts and were set down on the map, often prompting renewed speculations and fresh inventions. Faked maps were copied in whole or part by others who added still more mythical features, whether as the result of genuine error, excessive zeal or to further financial or political ambitions. Father Louis Hennepin – 'the greatest of geographical liars'[21] – made an assortment of spurious claims, mostly stolen from the equally fictitious reports of Sieur de la Salle whose map of middle and western America was highly inaccurate. Yet such fancies often found their way onto the official maps and were influential in determining subsequent explorations: 'The power of imagination over experience in the expansion and consolidation of geographical knowledge is exemplified [. . .] by the persistence of myths that constantly retreat into still unknown territory.'[22] This may be an inevitable factor in the settlement or colonization of other lands. The exercise of imagination is not a peripheral matter, but central to the process of exploration by which the unknown is brought into manageable form. Real exploratory initiatives and discoveries spring from preconceived images, however ludicrous they may later appear. Reports from the real world (as far as it can ever be divorced from conventional interpretations) are modified to fit the theory. Particular illusions and errors may be extremely persistent, continuing to provide the basis of the map onto which new information is forced.

The evidence of actual exploration has often carried insufficient weight to overcome faith in that which has already been mapped; at the same time, much real geographical knowledge has historically been unearthed by those chasing the myths. A classic example in the exploration of North America by the Europeans was the persistent myth of the existence of a northwest passage leading to Asia. Sebastian Munster's map of 1540, for example, is in no doubt. Accompanying a channel shown to pass through the continent are the words: 'This strait leads through to the Moluccas', a reference to the spice islands, difficult and dangerous to reach by land to the east, whose valuable trade explorers were keen to exploit. The error is thought to have been based on a combination of data from Verrazzano's journeys of 1522–24, on which the waters of Chesapeake Bay were mistaken for the Indian Ocean, and another explorer's navigation of the St Lawrence seaway and the Great Lakes.[23]

The illusion is repeated on Mercator's highly influential map of 1569, the one on which the great cartographer's continuing fame rests, using for the first time the projection that converted its loxodromes into the straight lines so useful for navigators. It was not of such value to Martin Frobisher who in 1576 took a copy on one of his many attempts to find the elusive northwesterly route to the east. Abraham Ortelius, another of the big names of sixteenth-century cartography, also followed the optimistic trend on the world map in his famed atlas, the *Theatrum Orbis Terrarum* (1570). Humphrey Gilbert's 1576 map placed the Moluccas temptingly just the other side of the passage: 'Despite the unsuccessful voyages of Frobisher in 1576–78, Davies in 1585–87, Hudson in 1607–8, Baffin in 1615–16 and others, hopes of a navigable strait remained.'[24]

An alternative American myth, again founded on the eagerness of explorers and traders to reach the orient once they found this huge new landmass blocking the sea passage they had sought, was that of a great westward-flowing river, the 'San Buenaventura', an ancestor perhaps of the elusive Bonaventure Hotel in Los Angeles. For decades explorers, travel writers and cartographers maintained that such a waterway existed, chasing it across the land from one blank space on the map to the next. No waterway was discovered, but one was eventually constructed. In 1903 United States support enabled Colombian secessionists to draw onto the map the line creating the new state of Panama, through which a man-made canal was driven. Another set of myths was applied to the vast plains at the heart of the North American continent, alternatively the Great American Desert, a barrier to further settlement, or a bounteous Garden. The emphasis shifted between the two at different points in the history of colonial expansion according to a variety of hopes and fears, needs and ideologies. Real patterns of settlement were affected. At first the dominance of the desert myth tended to discourage those in search of anything more than the rigours of a frontier wilderness experience. The myth of the garden, played on by railroad companies seeking backing for their transcontinental ventures, proved a powerful magnet to the eventual settlement of the plains. An equally persistent myth in the southern half of the continent was that of El Dorado, the fabulous city that entered geographical lore early in the sixteenth century: 'As late as the mid-nineteenth century, European map-makers represented the lake of gold, silver, and emeralds on their maps of South America, and governments continued to dispatch explorers in search of El Dorado; even some twentieth-century explorers have sought the lands of the Golden Man.'[25]

Maps such as these played an important part in the process of colonial control and settlement. Major Mitchell, surveyor-general of New South Wales from 1827 to 1855, either invented or exaggerated many geographical features, making landscapes appear more attractive to the prospective settler than was really the case. As Carter suggests, the settlers may have wanted to believe in this mythical version of the country. If the occupation of the territory was the goal the approach worked: 'From the point of view of colonization Mitchell's imaginary country was more real than any empirical one. It had, after all, provided a legend from which history could flow.'[26] The geographical equivalent of Linnaean taxonomy served a similar purpose. Divided up into blocks according to the imposition of an abstract grid, land could be sold off easily in lots. Buyers need not even visit the country before making their purchases. On the map the territory appeared neatly laid out according to rationally ordered principles, inviting their confidence and apparently making it clear precisely what they were getting for their money. Thus Charles Dickens has the title character of *Martin Chuzzlewit* (1843–4) buying a lot in the idyllic sounding American settlement of Eden on the basis of a map that shows the beginnings of a thriving settlement. On arrival he finds that he has been seduced by the map and is the owner of a piece of unimproved forest and swamp, a fate shared by many colonists. Maps aimed at the potential settler used pictures and symbols familiar from maps of the home country to give the landscape a feel more cosy and welcoming than the reality often turned out to be.[27]

Maps offered a means of both physical colonization and conceptual control. They imposed a form of differentiation and orientation on either plain or impenetrable forest, spaces that threatened to overwhelm. Initially many colonists and their promoters envisaged America as an Edenic paradise or a pastoral landscape promising an escape from the rigours of the Old World. This optimism was mixed or alternated with an opposing view of the New World as a harsh wilderness onto which it was necessary to impose some form of order. Along with idealized notions of the paradisal garden, the colonists brought from Europe a legacy of fear of the wilderness that was in some cases given a new and sharpened focus by the landscape with which they were confronted. From at least the Middle Ages wilderness had been seen by many as an objectification of evil, a chaotic and dangerous space. Those who strayed off the path and into such territories risked a fall into madness – literally, 'be-wildernment' – or sin, a threat dramatized in works ranging from the Bible to fairy tales such as Snow White and Little Red Riding Hood. For many of the early colonists, particularly the Puritans in New England, their settlements were fragile structures surrounded by a variety of unknown wilderness terrors. William Bradford, on his first landfall, described what he saw as 'a hideous and desolate wilderness, full of wild beasts and wild men [. . .].'[28] Earthquakes experienced in towns like East Haddam, Connecticut, in the seventeenth and eighteenth centuries were regarded as evidence of an abyss above which such islands of God and civilization were precariously balanced.[29] Four out of five of the supposed 'witches' accused at Salem lived beyond the bounds of the settlement itself, in a territory assumed to be that of Satan.

Images of the subjection and control of such unruly landscapes were widespread in European culture at the time of the settlement of the New World. Maps and perspective views such as Piero della Francesca's clinical representation of an *Ideal Town* were in vogue.[30] This was also the heyday of the formal Renaissance garden in which plants were regimented and their foliage trimmed into ornate patterns. Voyages to North America coincided with the publication of pastoral works such as Edmund Spenser's *The Shepheardes Calender* (1579) and Philip Sidney's *Arcadia* (1590). The pastoral ideal offered the possibility of a 'middle way', an ordered and controlled mode of existence bordered on the one side by the wilderness and on the other by the corruptions of civilization.[31] Transplanted onto the American landscape of the imagination, it was to have an influence on literature and politics lasting well beyond the period in which it might have seemed a practical possibility. As Alexander Wilson observes: 'The pastoral lawn, for example, not only predominates in suburban frontyards, but also stretches across golf courses, corporate headquarters, farmyards, school grounds, university campuses, sod farms, and highway verges.'[32] Tending to the pastoral garden in modern industrial societies might appear to be a digging down, literally, to a more real ground, the earth itself, as if it were not overlain by cultural strata. This is an illusion, of course, however enjoyable, and the domestic garden is an artificial product just as much as the house or street, bearing little relation to any natural foundation and highly commodified in the produce on sale at the modern garden centre. The garden offers an image of the shallowness of any cultural ground, a mere patch of earth artificially deposited but none the less capable of being naturalized and taken for something upon which to build empires great or small. More than any access to more authentic reality, the garden offers a precisely controllable, mappable space in which a multiplicity of clear boundaries, fences and hedges can be maintained; actual maps and models are offered in the gardening books for those in search of detailed guidance. Alternatively, the garden can be left to run wild, to go to seed, safe in the knowledge that any such chaos remains strictly bounded and contained.

Settlements planned according to the grid have been traced back at least as far as India in the third millennium BC.[33] It was spread widely during the conquests of Alexander and the Romans. The grid-plan was first imposed on the New World at Santo Domingo on a site initially chosen by Bartolomé Columbus, brother of the explorer. An image of its subsequent development in North America can be seen in a view of James Oglethorpe's 1734 plan for the settlement of Savannah, Georgia.[34] It shows the town taking shape on a parcel of cleared ground sharply delineated by the edge of a surrounding forest that stretches to the horizon on all sides apart from that of the river from which it took its name. Some of the settlement's buildings are already in place in a modified grid format. The rest of the grid-lot pattern is marked on the ground, waiting to be filled in. A wall is being built around the buildings further to stake out the town boundaries and to prevent the encroachment of the wilderness. Before long a similar grid would be applied to the entire country.

In the vast central and western parts of the United States the grid enabled newly appropriated lands to be packaged and sold before they had been adequately explored or surveyed. In 1784 Thomas Jefferson was appointed chairman of the congressional committee that had the task of devising plans for the organization of the western territories. He proposed that the land be divided according to a Ptolemaic grid based on lines of latitude and longitude. The grid would provide the boundaries of states that would be given appropriately synthetic names such as Polypotamia, Polisipia, Metropotamia and Assonisipa.[35] The geometrical neatness of Jefferson's scheme is not apparent when superimposed on the modern map, but it fitted Thomas Hutchins' inaccurate map of 1778. The details of Jefferson's plan were rejected, but the principle of division into a grid remained intact throughout the protracted debates that preceded and surrounded the great rectangular survey of midwestern lands that was eventually re-established, after slow beginnings, under a congressional act of 1796.[36] Huge tracts were divided up into small lots by the imposition of a grid structure. Lines of latitude and longitude marked many of the boundaries, including the 49th parallel between the United States and Canada. Initially attempts were made to survey the territory from the ground, but it proved slow and impractical. Regardless of topographical features, the landscape was divided eventually into townships six miles by six miles in size, subdivided into 36 sections of one square mile apiece. These were subsequently divided into smaller and smaller units as required, into half-sections, quarter-sections and even quarter-quarters, imposing the checkerboard appearance that remains clearly visible today. A similar rectangular system had been used by the Romans, but had taken greater account of local differences in terrain. In America the land was little more than a commodity:

> It had to be neatly parcelled for sale. Congressmen did not worry how to get from here to there, or precisely where South Pass was to be found, or whether a river ran this way or that. They worried most over two other problems: first, how to make as much as possible of the West quickly saleable, while spending as little as possible on costly surveying and mapping; second, how to package these parcels so that each purchaser would know precisely where his parcel was, and hence would feel assured that his title and ownership were safe. These overriding concerns, however irrelevant to the urgent problems of actual life in the West, shaped the patterns of American land ownership, land use, and land exhaustion for the following centuries.[37]

Issues such as the nature of landforms, the quality of soils and often most urgently the availability of water were not taken into account. These were the abiding realities for those who would live on the ground, but they were at this stage all but ignored: the abstract mapping came first. It left a legacy of problems for those who settled the land. However arbitrary, the map set the future agenda on numerous issues. Areas of low rainfall were divided up without regard to problems of irrigation. If the territory rather than the map had come first, borders might have been constructed with a view to the provision of equal access to precious sources of water. As it was the buyers of one lot might find themselves without water supplies that could have been included with only slight departure from the rigid framework of the grid. Where water was present the imposition of the grid pattern on the farming of sloping land created problems of soil erosion. Where this has since been countered by the use of contour ploughing techniques the patchwork remains visible beneath, the contour-sections themselves squeezed into the interstices of the grid.[38]

A twentieth-century image of the confrontation between wild and civilized can be found in constructions such as the interstate highways across the Louisiana swamplands north and west of New Orleans: ruler-straight concrete ramparts built on legs across a huge expanse of tangled wilderness neither solid nor liquid, a perfect metaphor of strict distinction between nature and culture. As Edmund Leach puts it: 'Visible, wild, Nature is a jumble of random curves; it contains no straight lines and few regular geometrical shapes of any kind. But the tamed, man-made world of Culture is full of straight lines, rectangles, triangles, circles, and so on.'[39] Another straight line is being inscribed onto the map in this comparison, of course, between the categories 'nature' and 'culture', especially when capitalized in this manner. What is described as 'nature' in fact only exists as such in being differentiated from 'culture': the line drawn is an arbitrary one, however significantly it may be experienced. The concept of wilderness does not usually have any meaning for those who have always lived within forests, deserts or other such landscapes. It exists only in opposition to the particular concept of civilization against which it is defined.

A dramatization of this opposition is found in Jenny Diski's novel *Rainforest* (1988). The protagonist, Mo, is a physical anthropologist carrying out fieldwork in the rainforest of Borneo. The forest is presented as a formless labyrinth, random and chaotic. Its growth is rampant and unstoppable, a threat to Mo's very existence: 'She could feel it pulling at her flesh as if it could not bear the contained boundaries of her body and were trying to find a way to unravel her.'[40] It is only through the imposition of an abstract grid that Mo can hope to comprehend the forest: 'in order to study the interaction of so complex an environment it is necessary to divide it into segments, to place over a small patch of forest a grid of squares, and to note exactly, day by day, everything that occurs within its boundaries.'[41] Mo has to keep her food, her clothing and, it seems, herself in sealed containers to protect them from the encroachments of the forest. Eventually it overcomes her defences and she is returned home to England to recuperate in a psychiatric hospital. To survive for any length of time in such environments, it seems, either of two approaches is necessary. One is to defeat the forest, to fell, clear or tame it on a much larger scale, as was the approach for many colonial settlers. The other is to apply a different kind of cartography, akin perhaps to that of the Mbuti pygmies, former inhabitants of the Ituri rainforest of Zaire studied in Colin Turnbull's anthropological classic, *The Forest People* (1961).

According to Mary Douglas, the Mbuti were able to 'move freely in an uncharted, unsystematized, unbounded social world'.[42] But this is an overstatement. The Mbuti lived according to their own mappings, even if they were more fluid than those imposed by colonial settlers. Confined on all sides by the enveloping forest, they had no idea of linear or any other kind of perspective that could account for the relative size of objects at a great distance. On a rare visit to open expanses of grassland, Turnbull's Mbuti ally was convinced that buffalo grazing several miles away were tiny insects. But the Mbuti had their own clearly established and coherent view of the world. As Turnbull puts it: 'In the forest life appears to be free and easy, happy-go-lucky, with a certain amount of perpetual disorder as a result. But in fact, beneath it all, there is order and reason; reaching everywhere is the firm, controlling hand of the forest itself.'[43] It was to the forest, which they experienced as the kind of living presence evoked by Diski, that the Mbuti prayed and from which their lives gained meaning and structure. Some rules were strictly enforced: the ultimate sin in a collective culture where hunting was carried out by the group was to seek only to better one's own catch. The Mbuti considered themselves to be an integral part of the forest organism, unlike Diski's protagonist who remains distant and separate from it. Her objectivism as a scientist renders her relationship to the forest as alienated as that of the avowedly objective Renaissance artist from the whole he sought scientifically to depict. Both attitudes are symptomatic of a single metaphysics that seeks to maintain strict boundaries between map and mapped.

Rather than being used to argue for the possibility of breaching all boundaries, a case such as that of the Mbuti should serve to demonstrate how excessively restrictive particular cultural frameworks may be, and to ask why. The non-forest environments described in *Rainforest* and in Diski's first novel, *Nothing Natural* (1987), are also worlds of defensive regimentation. The protagonist of *Rainforest* has a passion for order before she ever sets foot in the forest. In particularly fragile cultures like our own, Diski seems to suggest, such an attitude is necessary. Choice has been reduced to the alternatives of regimentation and a dissolution leading to madness. Little room is left for any more relaxed structure outside or between the poles of this opposition. Diski's narrative style itself imposes a rigorous framework. It is highly analytic, seeking explicitly to understand all that happens, to grasp each event within a tightly interlocking scheme. Strict control is maintained. Rather than simply confronting the reader with a series of disturbing events the narrative is rendered into a formal and theoretical pattern; it enacts precisely the kind of dilemma it describes. This is another way of guarding against the threat of chaos, a subject being explored theoretically in the text by Mo's friend Nick. To Mo's contention that chaos is simply chaos and cannot be thought, Nick replies that it can. His aim is to translate the abstract mathematics of chaos theory into an ordinary language in which it can more directly be conceived, something to which Mo, unsurprisingly, is opposed.

Diski's novel also embraces aspects of post-Einsteinian physics, a subject that has been held up in accounts of both the modern and the postmodern as an index of the breakdown of more certain Newtonian mappings of the universe. Einstein describes a world in which the conventional grid becomes useless for plotting the movement of bodies whose stability is thrown into question. In its place he offers a non-stable system of reference, an amorphous and slippery 'reference mollusc' the form of which is in constant change.[44] A world that embodies puzzling aspects of quantum mechanics and theories of entropy might seem to be further evidence of a modern/postmodern tendency towards the dissolution of structures and a drift towards entropy at a cultural level.[45] But gestures towards such theories may be just another way of seeking to contain any perceived dissolution (notwithstanding, here, the fallacy of mapping what happens at the sub-atomic level onto the cultural). Theorizing and narrating chaos and entropy may be one way of helping to keep them at bay. Nick is seeking not to accept decay and decline but to incorporate them through some kind of Hegelian *Aufhebung*, a negation that also entails a 'raising up' into a higher unity: 'A new structure emerges. Not chaos. A different structure.'[46] The map may change but it is not to be erased, just as Mo, to ward off her doubts, makes a last desperate attempt to move rather than to abandon her forest grids.

A correlation can be found between the acts of exploration and settlement and those entailed in the narration of the journey: as Carter suggests, they are two different but interrelated forms of 'setting down' in the landscape.[47] In being named and so called into being, the forms of the territory become a text within which orientation can be gained. Writing conforms to a similar dynamic, imposing its own linear order upon events. The particular way the Australian landscape was written by Western explorers was arbitrary but answered the specific requirements of colonization. We should also note the importance of a similar kind of narrative calling-into-being of the land in indigenous aboriginal culture. Aboriginal myths of creation speak of a world sung into existence by ancient totemic beings that wandered across the land in the legendary Dreamtime, leaving behind their traces in the forms of the landscape.[48] A narrative is thus created that renders the environment meaningful, negotiable and habitable. These meanings might be less exploitative than those imposed later by colonial explorers, but the difference is relative. In neither case is the meaning either inherent or objectively given. The landscape may thus be read as a palimpsest of different mappings.

A similar process of construction of meaning through narrative is depicted in *Waterland*. Rather than face up to apparent nothingness the people of the fens seek to construct something themselves out of the void. For some, such as the wealthy Atkinson family, this means the battle to claim and reclaim the land and to build business edifices that rise above the morass. These might share a symbolic function with other monumental structures standing out as cultural markers on landscapes that lack marked vertical features: the stones of Salisbury Plain, the ziggurat of Mesopotamia or the Egyptian pyramids, rigorously oriented according to cosmological belief.[49] For those of lesser means, such as the Cricks, the ancestors of Swift's narrator who live in the middle of the flatness, the only way to 'outwit reality'[50] is by telling stories, narrative yarns that offer both diversion and a form of meaning. 'Man', we are told, is 'a story-telling animal. Wherever he goes he wants to leave behind not a chaotic wake, not an empty space, but the comforting marker-buoys and trail-signs of stories.'[51] History is seen in the same terms, 'the Grand Narrative, the filler of vacuums, the dispeller of fears of the dark.'[52]

History and fiction, often seen as distinct, are far from entirely separable, like the maps and territories considered above. The point is not that historical events did not exist, like those made up in works of fiction, but that we can have access to them in the present only in already-textualized forms. History tends to ignore that which is not written down: 'history is essentially an act of interpretation, a re-reading of documents.'[53] But the interpenetration of fiction and history goes further than this. It is not simply that data are found in written records, but that even at the time of their comprehension and compilation the form and content of such data are shaped by an existing set of discursive practices. For E.L. Doctorow, whose own novels explicitly blend factual and fictional events and characters, 'there is no fiction or nonfiction as we commonly understand the distinction: there is only narrative.'[54] The historical text tends to employ narrative tropes in order to structure the record of events. It may, according to its originating context, present a vision of history as romance, tragedy or farce: 'The historical narrative thus mediates between the events reported in it and the generic plot-structures conventionally used in our culture to endow unfamiliar events and situations with meaning.'[55]

Narrative structures are widely used to make sense of experience, to impose an order on otherwise disjointed life. This applies to the real world as much as the pages of fiction. Life stories are shaped accounts that draw on elements of both mythic and real experience. Narrative or mythical frameworks are common in all societies, but they may be articulated with a particular emphasis in those subjected to rapid change. It is not a question of 'the crude weighing of "myth" against "reality".'[56] Myth is embedded in real experience, 'both growing from it and helping to shape its perception.'[57] Alternative historical narratives seek to challenge dominant myths and can be used to demonstrate their provisional status, although they are usually repressed. The same can be said of maps. If they are understood in this sense, as ideological constructs subject to contest, erasure or redrawing, we may be able to avoid the objectifying pitfalls of which Bourdieu warned. In Australia resistance might be found in the unofficial accounts of transported convicts:

> In dealing with authority, the convicts revealed its rhetorical foundations: maps and memos were instruments of strategy, not incontestable facts. Place names were figures of speech, places where one could speak. But this, after all, was how reason worked, by persuasion rather than by demonstration. For, *before there were facts, there had to be the fiction of facts.* There had to be agreement that space was not historical, that language was not metaphorical.[58]

But language is largely metaphorical, as is much more of human experience than is often admitted. Metaphor is often taken to refer to the poetic device or the rhetorical flourish in hyperbolic speech. Yet it is pervasive, both in language and the conceptual systems within which we organize thought and action. Metaphor is used both in everyday life and in scientific discourse.[59] The perspective that seeks rigorously to distinguish between map and territory is itself founded on a dominant metaphorical complex, what Deleuze and Guattari call an 'arborescent culture', the principal figure of which is the tree with its clearly differentiated hierarchies of root, trunk, branches and sub-branches. In its place they offer the decentred image of the rhizome, a complex, non-hierarchical system in which there is no absolute distinction between levels. The arborescent structure, they suggest, is that of a traditional representation or tracing of the real: 'The rhizome is something altogether different, a *map and not a tracing.*'[60] Rather than standing apart, as something separate and above, the kind of map these writers have in mind remains implicated: 'It becomes itself part of the rhizome. The map is open, connectable in all its dimensions, and capable of being dismantled; it is reversible, and susceptible to constant modification.'[61]

The distinction between the metaphorical and the literal

How might the rhizome influence the creation of a classification system?

can be dated back to Plato's dismissal of rhetoric or figural speech as the stuff of illusion and deception. Rather than seeking to demonstrate and prove, Plato argues, the sophist seeks to persuade through the hypnotic powers of rhetoric. But *all* language can be said to persuade, to argue for one perspective or another. Each language implies a metaphysics, a mapping of a particular view of the world, however covertly it might be expressed. Precisely such a metaphysics is itself implied in the traditional opposition between the metaphorical and the literal. That which is termed 'literal' is generally implied to be natural and immediate in its meaning. Its surface becomes opaque and the process of production by which such a meaning comes about is effaced. Nietzsche perhaps put it best when he argued that 'truths are illusions of which one has forgotten that they *are* illusions; worn-out metaphors which have become powerless to affect the senses; coins which have had their obverse effaced and are now no longer of account as coins but merely as metal.'[62] Philosophy, as Jacques Derrida has argued, cannot escape its own metaphorical bounds: 'If one wished to conceive and to class all the metaphorical possibilities of philosophy, one metaphor, at least, would always remain excluded, outside the system: the metaphor, at the very least, without which the concept of metaphor could not be constructed, or, to syncopate an entire chain of reasoning, the metaphor of metaphor.'[63] Obscuring this fact and seeing no farther than its own sedimented metaphorics, Western metaphysics has become 'the white mythology which reassembles and reflects the culture of the West: the white man takes his own mythology, Indo-European mythology, his own *logos,* that is, the *mythos* of his idiom, for the universal form of that he must still wish to call reason.'[64]

Everything is in a sense metaphorical, even if this is often not apparent or is deliberately ignored. The standard objection to this conclusion goes as follows: 'it cannot be that all the utterances in a language are metaphorical, since the very notion of metaphor only has a sense through its contrast with that of the literal.'[65] The above argument has already dealt mostly with this complaint. In tackling these questions we remain within the confines of a map from which we cannot entirely escape. The objection is true in a certain sense. The notion of metaphor does only exist as such in opposition to a notion of the

This, also, may be an interesting avenue to explore in terms of classification.

literal. This is Paul Ricoeur's position: that it was only with the development of a classificatory logic that sought to separate out the two that a specific and narrowed notion of metaphor came into use. That which is now dismissed as mere metaphor, a poetic kind of conceptualization, might have come first, as Giambattista Vico also argued in his *New Science* of 1744.[66] The classificatory logic that limited the domain of metaphor was the outcome of a process akin to that of metaphor itself: a grouping together according to similarities: 'A family resemblance first brings individuals together before the rule of logical class dominates them. Metaphor, a figure of speech, presents in an *open* fashion, by means of a conflict between identity and difference, the process that, in a covert manner, generates semantic grids by fusion of differences *into* identity.'[67] The underlying logic of the two is thus the same.

In seeking to undermine the opposition between the metaphorical and the literal it is important not to dismiss the *notion* of the literal. It is a notion that carries great weight in Western culture. The blatant propaganda maps described in Chapter 2 are like figural or rhetorical language, used deliberately to further an argument rather than to attempt a simple statement of truth. 'Literal' language is like the ordinary map: each appears to be a neutral representation. But no such neutrality is possible in either case. Each implies a view of the world, a bias that may be all the more insidious for its claim to objectivity or transparency. Like maps, metaphors can create realities rather than just express existing ones. As George Lakoff and Mark Johnson suggest: 'It is reasonable enough to assume that words alone don't change reality. But changes in our conceptual system do change what is real for us and affect how we perceive the world and act upon those perceptions.'[68] Metaphors establish mappings according to which future interventions are shaped in potent acts of self-fulfilling prophecy.

5 The Thicket of Unreality

[. . .] we have used our wealth, our literacy, our technology, and our progress, to create the thicket of unreality that stands between us and the facts of life.

Daniel Boorstin[1]

The most serious crisis in the history of mankind [. . .] *turned on a question of appearances. The world came close to total destruction over a matter of prestige.*

Stephen Ambrose[2]

Concern about the potency of images as against the reality they were thought to pervert or obscure, rather than accurately to map, became acute during the 1950s and 1960s in the West, particularly in the United States. Some years before the term 'postmodern' gained any wide currency the growth of media such as television and advertising seemed to some commentators to result in the swamping of a real that became lost beneath a wave of image, fantasy and illusion. In describing this process writers such as Daniel Boorstin and Vance Packard adopted a tone at times close to moral outrage. If the line between the image and reality or that between reality and representation was being blurred, it was, wherever possible, to be restored and those responsible were to be castigated. For Packard it was the cynical 'hidden persuaders', the advertising men and women of Madison Avenue and their allies in the social sciences, who were to be admonished for their devious manipulations. From Boorstin came criticism of the purveyors of 'pseudo events' masquerading as news, of vacuous celebrity, and, everywhere, those who were submerging any authentic reality beneath a gloss of ersatz imagery. Where later observers like Baudrillard are happy to accept the merging together of image and reality these writers seek to guard the sanctity of a distinction between the two.

Packard's *The Hidden Persuaders* (1957) abounds with cases from the world of advertising in which images, illusions or fictions of one kind or another seem to outweigh simple material facts. Consumers tend to imbibe the image of a product as much or often more than the specific qualities of the product itself. Rather than following along behind the reality of the product the image may lead the way, determining how the product is experienced. As the research director of one New York advertising agency puts it: 'People have a terrific loyalty to their brand of cigarette and yet in tests cannot tell it from other brands. They are smoking an image completely.'[3] The commodity, for Guy Debord, is a 'factually real illusion [. . .]'.[4] The spectacle, as he terms the spread of this phenomenon, 'is not a supplement to the real world, an additional decoration. It is the heart of the unrealism of the real society';[5] 'Everything that was directly lived has moved away into a representation.'[6] Indeed, the spectacle 'is the map of this new world, a map which exactly covers its territory.'[7] As we saw in Chapter 1, this process is often associated with the growing hegemony of television images in events ranging from the war in the Gulf to the Los Angeles riots of April 1992 or the O.J. Simpson affair.

A loss of the 'directly lived' into a realm of representation or spectacular is what concerns Boorstin in *The Image – A Guide to Pseudo Events in America* (1961), a work that typifies a more widespread complex. For Boorstin: 'The American citizen lives in a world where fantasy is more real than reality, where the image has more dignity than its original.'[8] The greatest threat to modern American life is the menace of unreality:

> We risk being the first people in history to have been able to make their illusions so vivid, so persuasive, so 'realistic' that they can live in them. We are the most illusioned people on earth. Yet we dare not become disillusioned, because our illusions are the very house in which we live; they are our news, our heroes, our adventure, our forms of art, our very experience.[9]

But disillusioned we must become, suggests Boorstin, the explorer of this increasingly fictionalized world armed with the rational machete with which he hopes to cut his way through the 'thicket of unreality'. A similar drive seems to underlie Packard's study. Although he pays due heed to the more concealed motivational factors unearthed by the psychiatrists, sociologists and anthropologists used by the advertising industry, their probings and manipulations are characterized as regressive for man 'in his long struggle to become a rational and self-guiding being'.[10]

In his analysis of news coverage Boorstin bewails the flooding of our television screens and newspapers with accounts of 'pseudo events' rather than the reality of a real outside world. He may be right to criticize the manipulative and superficial nature of much of what passes for news, but he is wrong to assume the existence of some kind of golden age in the past when this was not the case, when, in this example, it was a question purely of news-gathering rather than news-making. There never has been any such thing as news-gathering, in the implied sense of a merely passive and neutral recording of events 'out there' in the world. All news is constructed to a greater or lesser degree, according to a host of contingent and often arbitrary factors which determine which events are defined as news and which are not, when, how and by whom. If there has been an increase in the volume of news that is more overtly constructed it can be only a quantitative, not a qualitative, shift.

The same goes for the political sphere, which Boorstin sees as overtaken by 'pseudo events' such as media interviews or speeches staged artificially for the media rather than the live audience. Politics has indeed become highly mediated and image-based, a process often developed to considerable levels of sophistication. In election campaigns party managers devote much of their effort to controlling the daily quota of images supplied to the media; campaign events look more towards the photo-opportunities they will supply than to direct contact with the untamed voter. In Britain this approach was pioneered by the Conservatives but it has also been embraced by the Labour Party.[11] We should question the extent to which there ever was any purely unmediated political process, however, any rational political debate or policy-making freed from rhetoric, ideology and symbolism on any side. When John Major affected to return to such a realm in his adoption of a 'soapbox' platform in the 1992 general election campaign its purpose remained primarily symbolic and geared to the images it would furnish, a rhetorical claim to a notion of direct engagement rather than the real thing.

The pre-packaged experience of the tourist is said to have replaced authentic forms of travel that involve real contact with different places. The territory to be visited is mapped in advance, whether in the simulacra of specially designated tourist sites/sights, pre-programmed itineraries or actual maps of routes to be followed at micro- and macrocosmic scales. The stay-at-home holiday promised by virtual reality technology might be little more than a logical extension of the trend. Computer-generated beaches, safaris or Aztec ruins offer the ultimate in sanitized and controlled tourist experiences, travel freed from all the difficulties of actual contact.[12] Tourists are too easily satisfied by such contrivances, Boorstin suggests. It might better be argued that tourists often go in search of precisely the kind of authenticity he champions. It is this search for the authentic, paradoxically, that leads into the fake, the contrived or the inauthentic. Tourists often seek access to the 'back' areas of the sights they visit, the behind-the-scenes reality masked by the staged display up front.[13] But the back itself is often staged in a way that undermines the opposition. Institutions such as NASA and the FBI, to name but two, allow visitors not just to see informational displays but to view some of the inner workings. The visitor to the FBI headquarters in Washington, a popular although unlikely sounding tourist attraction, can peer though glass panels to watch forensic staff at work or to see the vast firearms collection. But the tour remains almost as much a show as Disneyland. It presents a superficial gloss, seeking to maintain the heroic myth of the G-man, that gains credibility from the apparent glimpse behind the scenes. Needless to say no mention is made of the more shady dealings of the bureau, just as the NASA exhibits at the Kennedy Space Center in Houston make no mention of disasters such as the 1967 launchpad fire or the loss of the shuttle *Challenger*.

The selective opening of such spaces to the public involves no more than a *rhetoric* of openness. Washington abounds in examples, including the admission of visitors to Congress, Senate and the Supreme Court. In Britain the doubtful or fearful are invited to tour the Sellafield nuclear plant. The fact that admission is granted implies that there is nothing to hide, as if what can be seen is any real measure of safety or any guarantee of democracy. Parliaments and congresses, and even cabinets and committees, supposedly the back areas where the business of government really occurs, are in many respects fronts. The real exercise of power and decision-making usually goes on elsewhere and out of sight. Architecturally this rhetoric of openness is found in edifices such as the Lloyd's insurance building in London. The functional ducts and pipework are displayed on the outside in an apparent celebration of transparency that, a cynic might argue, implicitly disavows the underlying and concealed manipulation of the economy undertaken by the City for which it has become a postmodern symbol.

In *The Tourist: A New Guide to the Leisure Class* (1976), Dean MacCannell concludes, hardly surprisingly, that the experience of the tourist would be more superficial than that of an ethnologist engaged in fieldwork in the same place. MacCannell concedes that an element of mystification might remain in the experience of the ethnologist, but the basic opposition he establishes is a false one. If there are limits to the perspective of the tourist or ethnologist the differences are relative rather than absolute. The tourist may obey a cruder map than that used by the ethnologist, just as the explorations of the ethnologist may be guided by a plan incapable of depicting many of the beaten tracks used by the subject under study. But the subject is also immersed in the frameworks of a cultural cartography. No individual or social group exists in a state of transparent or unmapped immediacy. We all occupy taken-for-granted mappings and assumptions that, as Bourdieu puts it, 'are the very house in which we live.' We cannot move out into some kind of open or transparent space; the best we can hope for if unsatisfied is to move house or rearrange the furniture according to a different map. The shortcoming of MacCannell's argument can be traced to the broader framework in which it is placed. The closure of social spaces, he suggests, is a function of modernity. Against this he posits an open form of 'primitive' society, an opposition that is not tenable. However much twentieth-century industrial society may have multiplied the number of formal institutions and media interventions involved in our lives, and however much we might perhaps be tempted by the idea of a technologically simpler lifestyle, it is wrong to assert the existence of an earlier moment of immediacy or transparent openness. Cultural mappings remain necessarily opaque.

As far as the tourist sight is concerned the reproduction of its image is an important factor in setting the tourist off in search of the real object. The essence of the phenomenon might be found in precisely what Boorstin opposes, the process by which the image comes before rather than after the experience of the real thing. The sight becomes authenticated only when its copy is produced. The tourist map also plays its part in the process: 'This kind of map designates places where a ravenous consumption picks over the last remnants of nature and of the past in search of whatever nourishment may be obtained from the *signs* of anything historical or original. If the maps and guides are to be believed, a veritable feast of authenticity awaits the tourist.'[14]

Boorstin mourns what he sees as the displacement of the original work of art by its more widely distributed copy, not entertaining the possibility that the status of the original itself might be open to question. For Walter Benjamin the loss of the aura of authenticity in the artistic work is a radicalizing change, removing it from a dimension of ritual contemplation to one of political practice.[15] But he seems to share with Boorstin the presupposition that there existed an earlier period, in this case before 'the age of mechanical reproduction', in which we could speak of a properly authentic original. The dominant conception of the work of art as an original creation – usually implying the existence of an identifiable author – is a relatively recent invention. Many great works are collective enterprises in which the canonized artist drew on and adapted a range of existing material. However brilliant or pleasing they may be, works of art must be seen in wider contexts. Few can be described as original, in the stronger sense of the term, as standing in an elevated position on their own. It is often difficult to trace the 'original' among a series of versions completed by the artist. Works are also dependent on the techniques and technologies that happen to be available at the time and the canon of others on which they draw or against which they react. Any work has also to be seen in the much broader context of the social environment in which it is produced.[16]

Daniel Boorstin is not alone in reacting so negatively to the postwar blurring of certain boundaries, both physical and conceptual. Other commentators such as Evelyn Waugh and Nancy Mitford had already sought to reimpose strict social mappings onto a territory blurred by the upheavals brought by war and its aftermath. For the likes of Waugh and Mitford (and more radical critics such as George Orwell and Richard Hoggart), it was a question of resisting shifts in 'the cartography of taste' that threatened to render the previously differentiated and hierarchical world 'flat as a map'.[17] The spectre of Americanization, of widespread cultural homogenization, was evoked: 'The cartography of taste here took on an aggressively reactionary flavour. The codification of the system of U and non-U was designed to clarify and redefine boundaries rendered opaque by post-war "affluence".'[18] Notions of authenticity were articulated to distinguish new and/or American forms such as rock-and-roll or rhythm-and-blues from 'natural' blues, folk music or traditional jazz.

The 'thicket of unreality' is not one from which we can ever emerge. Nor is it something we ever entered in the first place, in the sense that there was some earlier stage of existence in which map could be separated from territory. The thicket may in places be thornier than we would like, but in such cases our efforts should be devoted to making it less prickly rather than expecting to emerge into some kind of daylight clearing of purely rational immediacy and self-presence. Questions of taste involve socio-cultural imperatives rather than simple matters of individual like or dislike, however much that is how it seems to us at the time. When we buy or imbibe particular products we signify our membership of particular social groups or subgroups of one kind or another.[19] Advertisements offer membership of product communities, broad or exclusive. Consumption is a collective act, a system of meanings and values, however debased it might appear in some of its worst manifestations. It is not a matter of individual choice and expression or the fulfilment of natural needs.

The supposed practical utility or use-value of objects is often little more than an alibi for their role in the process of social signification or 'symbolic exchange value' as Baudrillard terms it. The conspicuous consumption and wastage of the rich serves to demonstrate social status. Similar dynamics can also be found among the less wealthy who aspire, consciously or not, either to greater things or simply to live up to the conventional norm. Where material wealth is scarce, deductions for the purpose of waste or sacrifice may be made first, to meet socio-cultural rather than physiological imperatives, and not just after those made to meet physical needs. Thus the status-conscious wife in the film *A Private Function* (1985), set during the postwar years of austerity and food shortages, pours into her garbage the contents of a full tin of corned beef and a tin of peaches. This may seem to be a futile waste of much needed food. Status can also be nourishing, however. The value of the food as conspicuous waste, apparent evidence of belonging to a superior social group, is greater than its edible worth: 'I'm not having them think we put just rubbish in the bin.' We should not underestimate the potency of cultural imperatives that can outweigh the need to meet even such apparently fundamental and 'natural' needs as those for food and shelter.[20]

The very establishment of modern America as a national entity whose diverse peoples share a common daily texture of life was to a large extent founded on the creation of communities of people defined by the products they consumed, as Boorstin himself suggests in the third volume of his history of the United States. These communities were shallower and more superficial than those held together by ties of religion and tradition or shared experiences such as voyaging across ocean and prairie. But they were able to draw together a wider cross-section of the population and to offer a common national garb into which many new arrivals could slip with relative ease. National chains of shops displaced local stores; franchised establishments permitted Americans to enjoy the same standardized brand-name products anywhere in the land, while mail-order catalogues and the provision of a free postal delivery service catered for those living in isolated rural areas, often at the cost of local community ties. 'In a word, it was a defeat of the seen, the nearby, the familiar by the everywhere community.'[21] The tone here is strangely approving given that the dynamic described, a displacement of supposedly immediate reality by a thinner more mediated ground of consumerism, is essentially the same as that criticized in *The Image*. The impression given is that the loss of a more authentic face-to-face dimension was at this stage a price worth paying for what Boorstin sees as the democratic openness of the new consumption communities. He makes little mention of the plight of those who could not afford the goods newly displayed to all in department store, catalogue and advertising copy, and offered as the sole means of admission to these new forms of communal belonging.

For Baudrillard, the spread of the image has reached a point where the real itself no longer exists. It is not something that remains ultimately recoverable in any way beneath its image or representation. This view is perhaps best expressed in his classic reading of Disneyland:

> Disneyland is there to conceal the fact that it is the 'real' country, all of 'real' America which is Disneyland [. . .]. Disneyland is presented as imaginary in order to make us believe that the rest is real, when in fact all of Los Angeles and the America surrounding it are no longer real, but of the order of the hyperreal and of simulation. It is no longer a question of the false representation of reality (ideology), but of concealing the fact that the real is no longer real, and thus of saving the reality principle.[22]

Baudrillard draws up a schema of four successive phases that he says have led to this evaporation of the real. Initially the image reflected a basic reality. This would accord with everyday Western understandings of the relationship between reality and its image or representation. The reality is the dominant instance, the image merely something secondary that produces a more or less exact copy of it, the map simply a product of the territory. In the second phase the image comes to mask and pervert a basic reality. This would be the situation as described by Boorstin, where a fallacious supremacy has been achieved by the image over a reality that remains sovereign underneath, if only the veil of illusion and layers of cartography could be torn aside. The third phase comes when the image masks the fact that there is no underlying reality, no virgin territory beneath the map. This leads to the current phase, in which the map is freed from any notion of distinct territory, when the image 'bears no relation to any reality whatsoever: it is its own pure simulacrum.'[23]

It is not always easy to separate out the realm of images from that of the real in the fabric of contemporary life. Much of the reality we inhabit is comprised of imagery of one kind or another, as we saw in Chapter 1. Efforts are often made to recover apparently lost elements of reality or authenticity. 'When the real is no longer what it used to be, nostalgia assumes its full meaning. There is a proliferation of myths of origin and signs of reality; of second-hand truth, objectivity and authenticity.'[24] This nostalgic search for authenticity may prove counter-productive. America is a land obsessed with realism, suggests Umberto Eco. If a reconstruction is to be credible 'it must be absolutely iconic, a perfect likeness, a "real" copy of the reality being represented.'[25] In the search for reality what is found or created is only the ever more convincing fake: 'To speak of things that one wants to connote as real, these things must seem real. The "completely real" becomes identified with the "completely fake". Absolute unreality is offered as real presence.'[26] This kind of dialectic between real and fake, authentic and copy, is far from new. A similar tension was felt in the second half of the nineteenth century, for example, when the new possibilities for imitation and illusion created by technological developments in the fabrication of materials provoked demands for a return to a real or an authentic believed to be threatened with displacement.[27]

The search for solid foundations only leads us further into a labyrinth in which the real and the unreal, the territory and the map, become more difficult to distinguish. Where traditions are absent or threatened they are invented.[28] A classic example is the British monarchy. Often dismissed from all sides of the political spectrum as of no real consequence, the monarchy may as Tom Nairn suggests play an important unspoken role in the construction of a simulacrum of nationality.[29] This is not the peripheral matter to the national life that it might seem but 'an essential factor in the standard identity-kit of that existence itself. It is a vital part of "who we are" [. . .].'[30] Nations are relatively arbitrary constructs requiring symbolic reinforcement and maintenance, as was seen in Chapters 2 and 3. One important feature of this process is the map of the territory, but it also involves collective myths that establish important cultural cartographies. Our mythic notion of the country is not a mere fantasy: 'One can think of it rather as a set of mental map-survey points implanted in the communal psyche. These are not necessarily referred to every day but remain indispensable for knowing where one is (especially when lost, or at moments of crisis).'[31]

Royalty offers a set of powerful totemic symbolisms and identifications, the more so as it loses direct political influence and so can affect to stand above sectional or class interests. Various supposedly ancient royal traditions were deliberately constructed, invented or resurrected from irrelevant obscurity at times when the national self-image was in need of polish. Examples include the investitures of the future Edward VIII as

Prince of Wales in 1911 and of Prince Charles in 1969. Critics who dismiss the influence of the monarchy may miss the point. As Nairn puts it: 'It could be argued of course that the Royal conservative-national identity was fraudulent, that a *real* Britain was betrayed and disguised by it [. . .].'[32] But this would wrongly suppose that there were some pre-given or essential Britain that had meaningful existence outside any such fabrications. The national entity constructed with the aid of royal symbolism in the late eighteenth century may in one sense have been pure fiction, but rationalist dismissals fail to appreciate the potency such fictions can have: 'What they ignore is the powerful self-fulfilling element built into such conduct: the extent to which "belonging" and social sentiment attach themselves to some emblem itself renders the magic "real", and threats to it correspondingly serious.'[33] The result is that fantasy becomes a part of political reality. The monarchy 'is not decorative icing on the socio-political cake. It is an important ingredient of the whole mixture [. . .].'[34]

Other ideological effects may also result. The privilege and inequality embodied by royalty may reflect the real inequalities of British society. Yet by flaunting them in this particular way their reality can be denied in a curious kind of double-bluff. Any suspicion that the displayed privileges of royalty may reflect inequalities in the rest of society 'is disavowed by the fixed, deeply reassuring conviction that all that is mere appearance'[35] rather than reality. The terms in which the political character of the nation are discussed are thus shifted onto a ground that ignores the kind of issues that might be raised by a more radical analysis of the capitalist state. The parliamentary arena (the political real when counterposed to royal show) is in fact more fittingly described as a show that effectively masks the realities of power located in entirely unrepresentative spheres of both state and economy.

In Baudrillard's account the phenomena of the hyperreal or the simulacrum are associated with a series of particular developments resulting from the establishment of a society founded on modern consumerism. He turns out to have more than might be expected in common with Boorstin, retaining a faith in the existence of an earlier period in which the dominance of reality over the image was a fact, not just in terms of how it was understood at the time but in terms of the reality itself. In the quotation from which I have drawn the title of my first chapter – 'Henceforth, it is the map that precedes the territory' – it is the 'henceforth' that I wish particularly to question. Baudrillard's argument is restricted to a particular context that he later terms the postmodern. A more radical analysis is required in which the map is found always to have been implicated in the territory. The differences of the two positions can be seen in two contrasting readings of *The Right Stuff*, both the novel by Tom Wolfe (1981) and the film written and directed by Philip Kaufman (1983). The subject of *The Right Stuff*, the early days of the space-race between the United States and the Soviet Union, also suggests the power of symbolic matters in international relations and the potency with which a Cold War map was imposed on the world.

The Right Stuff is structured around an opposition between the Mercury Seven astronauts, America's first men in space, and the world of their forerunners, the jet test-pilots from whose ranks some of their number were drawn. The astronauts became celebrities before setting a foot near the launch pad. Announced to the world in a blaze of publicity, they were instant media stars. Their status fitted in part with Boorstin's definition of the celebrity as 'a person who is known for his well-knownness' rather than for any other achievement.[36] The seven astronauts were initially famous simply for being famous, for their appearances on radio, television and in the newspapers, and for the huge fees paid by *Life* magazine for their exclusive ghosted stories.

The case of the astronauts seems to blur the boundary between hero and celebrity so carefully maintained by Boorstin in his lament of the apparent eclipse of the former by the latter. The seven are more than celebrities in the usual sense. Although the media fanfare pre-empts the event and their ultimate role is limited they still have at some point to face up to elemental forces, rather than merely braving the studio of a television chat show. They do occupy an entirely mediated environment, however, in terms both of the enveloping media attention and the complex technological mediation of contemporary computer-controlled rocketry. This absorption within the technology is explicitly thematized by the text. Much of the energy of the seven is devoted to an attempt to establish their credentials as real pilots of their craft, with some genuine hands-on control, rather than being relegated to the status of mere 'redundant components' or 'Spam in a can' as Chuck Yeager, the definitive heroic test-pilot, puts it. While other test-pilots are pushing the new X-15 rocket plane ever closer to the 50-mile high frontier at which space begins the astronauts remain little more than test specimens.

The experience of Yeager and his colleagues, both book and film suggest, is more authentic than that of the astronauts. Jet technology may be complex, but there is still a central role for the pilot. Nobody suggests that Yeager might be in competition with a chimpanzee, as is the case with the astronauts. It is a trained chimpanzee that makes the first space flights, both sub-orbital and orbital. Through the eyes of the test pilot, Wolfe describes the sensation of taking off at dawn in an F-100 fighter, hurtling into the air 'so suddenly that you felt not like a bird but like a trajectory, yet with full control, full control of *five tons* of thrust, all of which flowed from your will and through your fingertips, with the huge engine beneath you, so close that it was as if you were riding it bareback [. . .].'[37] The test pilot also has room for improvization. When Yeager falls off his horse, injuring his ribs on the eve of an attempt to break the sound barrier, he finds a makeshift way of locking down the door of his experimental X-1 jet and goes on to make history. The case of the astronauts offers a telling parallel. The construction of their heroic image is essential to the continued funding of the multi-billion-dollar project. The astronauts play on this notion of heroic individual agency to make demands of the programme controllers. They argue successfully for the installation of not just a window but also an escape hatch with explosive bolts that can be operated by the occupant himself. Unfortunately the only use of the hatch we see is by Gus Grissom when he loses his capsule after appearing to panic and open the hatch in the sea after splashdown. For the test pilot the human-controlled dimension is one that allows a creative, resourceful inventiveness to overcome adversity. For the astronaut, cocooned within a much deeper technological web, it offers the chance to 'screw the pooch', as the lingo has it.

Media coverage continued to emphasize the active role of the astronaut, almost in direct proportion to its absence. In so doing, it highlighted one of the central contradictions of the contemporary celebration of new technology. The wonders of science promised a world of effortless and automated control. But advertisers of new gadget-ridden products such as cars also played on the opposed desire of many consumers for an assertion of their own powers of agency, a desire particularly acute in the social context of a perceived threat from the increasing bureaucratization (and, supposedly, feminization) of life. Advertisements offered sublimated fulfilment through the apparent mastery of technology involved in understanding the latest model to roll off the production line. New gadgets might do much of the work themselves, but they created their own arcane domains to be controlled. The proclaimed agency of the astronaut offered a similar satisfaction at a collective level.[38]

Yeager's successful breach of the sound barrier can be seen as a pivotal moment in *The Right Stuff*. The event does not at the time become news, but is kept secret. The momentous achievement could have been flashed around the world in triumph. But the enthusiasm of the press officer on the scene is overruled by the security men. At this stage 'real' concerns of military advantage and national security are deemed more important than symbolic flourishes. A different attitude follows the launch of the sputniks and the flight of Yuri Gagarin. American prestige is damaged and considerations of image come to the forefront. America is desperate to achieve a response, even risking lives in the efforts to get something up 'quick and dirty' as an immediate response. The humiliating failures of the initial rocket tests are as a result conducted in full public view on live television.

While the astronauts bask in a constant glare of publicity, the test-pilots occupy a closed world, isolated on an Air Force base in the high desert of California. It is a world of stoic rituals and totemic customs. One female groupie demonstrates her lack of inside knowledge by asking why a particular pilot has not been immortalized by having his photograph mounted along with others on the wall of the bar. The pictures are of those whose aircraft have come to grief. This is another telling point of contrast. For the test-pilots, iconic representation comes primarily after the event, at the end of their careers and as a testament to what they have done and the fate they have met. How different this seems to the endless media representation of the Mercury Seven, much of it *before* their work has begun.

It is not only the image that seems to precede the eventual

flight of the astronauts. So does the experience itself. For Alan Shepard, the first to go into sub-orbital space: 'His launching was an utterly novel event in American history, and yet he could feel none of its novelty.'[39] He has experienced it all in advance in detailed simulation and the flight itself is almost an anticlimax: '*It was all milder! easier!*'[40] The supposedly fantastic view from the capsule is not as impressive as those to which he had been conditioned by photographs flashed onto a screen in the flight simulator: 'The real thing didn't measure up. It was not realistic.'[41] By the time John Glenn made the first orbital flight: 'No man had ever lived an event so completely ahead of time.'[42] Even when sitting atop the rocket, then, or in a flight of some danger however little control he might have, the astronaut has little experience that has not exhaustively been preconditioned or preceded by a welter of imagery. When such an experience does occur it becomes a newsworthy event in its own right, such as the rescue of an escaped satellite by the crew of the shuttle *Endeavor* in May 1992. Hands-on control was established, literally, as four astronauts grappled with the satellite in an unrehearsed space-walk manoeuvre forced by the failure of the planned strategy. The mission became a live television spectacle. Grainy video pictures were transposed onto newspaper front pages, projecting an image of heroic endeavour gratefully received in a nation whose previous broadcast spectacular had been the Los Angeles uprising two weeks earlier.

The Mercury project was caught up in matters of image and symbolism on a broader canvas. It was, after all, a race against the greater initial success of the Soviet Union. Rival developments in rocketry had serious implications for two superpowers equipped with nuclear weapons in need of new delivery systems. The launch of the sputniks was not a revolutionary feat of engineering at the time, however. Both superpowers had gradually been moving towards such a capability since the end of the war. It was the symbolic rather than material implications that gave the event such dramatic impact.[43] Faced with what many believed to be the incipient threat of Soviet expansion across the globe, Americans had drawn comfort from the belief that the United States remained technologically superior. The presence of a Soviet satellite circling overhead seemed to undermine such an assumption. American prestige was dented.

The reaction was at times close to panic. Leading newspapers and magazines ran editorials asking searching questions. Had the nation at large, and the Eisenhower administration in particular, been complacent? What would be the impact on the battle for the hearts and minds of the rest of the world? This was an important question, one that seems central to an understanding of the impetus towards what became an all-out race in space. The period was one of colonial retreat, in which new independent nation-states were emerging onto the map in several parts of the world. Whether for defensive or expansionary reasons, both superpowers had an interest in the current and future allegiances of these states. Capitalism and communism were seen as two rival models of progress, each seeking to present itself as the most appealing to others. Achievements in space were seen by both sides as offering an impressive demonstration of status. The Soviet Union could claim within 40 years to have come from the backward czarist society of 1917 to global technological superiority. Such an image might be attractive to countries themselves only beginning to cast off the shackles of imperialism, especially at a time when America's avowed postwar opposition to colonialism had been undercut by its support for imperial holdings such as those of France in Vietnam.

The frenzied opening of the space age produced 'a new symbolism that had more reality for mass politics than did the actual data on *Sputnik*, Soviet and American defence budgets, missile progress, and science and education.'[44] The importance of this symbolism was recognized by the Soviet leadership. To the writer of an editorial in the Moscow journal *International Affairs*: 'The victory in space is the triumph of Socialism and Communism – the new social system [. . .] has demonstrated its undisputed superiority over the society based on competition and human aspiration.'[45] In fact lagging behind the United States in all but the development of large boosters, the Khrushchev administration was happy to encourage the myth of its technological and military superiority: 'the Soviets leaned on promise and appearance, not on reality – it was more important for the USSR to seem invincible and progressive than for it to be so.'[46]

The American military lobby was equally well served by illusions about the meaning of *Sputnik*, which provided the political leverage for its own expansion. It was a time when 'an image of technical dynamism was as important as actual weapons' and when 'prestige and perceptions were as important as actual military force.'[47] This had become true of American Cold War policy in general. A commitment was made to Korea despite the judgement of the Joint Chiefs of Staff that the peninsula was of little importance from the standpoint of military security. The symbolic value of Korea was often emphasized ahead of other strategic dimensions, particularly once American troops had been deployed in 1945 and again in 1950, although the Korean war also proved useful to those who sought to justify a broader remilitarization of American foreign policy. The loss of Korea to Soviet influence or an independent socialist regime would be a blow to American prestige. It might cast doubt on the depth of US commitments elsewhere. A increasing emphasis was placed on the importance of appearances and credibility, as opposed to actual capabilities. This approach was enshrined in a key policy statement by the National Security Council, NSC 68. The new strategic map blurred the distinction between vital and peripheral interests that some foreign policymakers had sought to maintain when allocating limited resources. For the authors of NSC 68, *any* setback to the United States could be equally damaging. American credibility could be at risk even where little was at stake on the military or economic fronts. 'The implications were startling. World order, and with it American security, had come to depend as much on *perceptions* of the balance of power as on what that balance actually was.'[48] The number of apparent threats multiplied as a result, each reverse – the 'loss' of Cuba in 1959, the failure of the Bay of Pigs invasion in 1961 – seeming to increase the credibility stakes in the next.

The Cuban missile crisis of 1962 saw an escalation of tension out of proportion to the material realities at stake. The extent to which the strategic balance was altered by the deployment of Soviet missiles in Cuba, or the removal of American missiles from Turkey demanded by Khrushchev, remains subject to debate. The precise nuclear arithmetic was not the central issue for the Kennedy administration. Kennedy's principal concern was with the symbolic dimension. Whatever their military significance, the missiles were seen as a threat to American prestige abroad and that of the Kennedy presidency at home.

Kennedy insisted on seeing Khrushchev's policy as a test of his personal resolve. This was despite America's overwhelming nuclear superiority and provocative moves such as the deployment of Jupiter missiles close to the Soviet border in Turkey and aggressive campaigns against Cuba that included military exercises based on preparations for a possible invasion. Khrushchev was also reluctant to back down and lose prestige. The world came to the brink of nuclear war on a matter of image as much as substance. The American intervention in Vietnam was also propelled in part by questions of appearance. Vietnam was to be the arena in which American resolve finally would be proved. Withdrawal thus became extremely difficult for a generation of policymakers, even after the inevitability of defeat was accepted. A number of post-Vietnam foreign policy interventions can also be seen as symbolic strikes designed to give a boost to the American image rather than as measured responses to real threats: the more obvious examples include the invasion of Grenada in 1983, an apparent demonstration of vigour just weeks after the humiliating loss of 241 US marines killed by a car bomb in the Lebanon.

It was only when he was convinced of the symbolic importance of the project in superpower competition that President Kennedy put his weight behind the expansion required for a project to go to the moon. Eisenhower had been reluctant to commit huge resources to a programme justified primarily in terms of prestige.[49] The nature of the enterprise was encapsulated in a memo approved by Kennedy from James Webb, the NASA administrator, and the defence secretary Robert McNamara:

> Dramatic achievements in space [. . .] symbolize the technological power and organizing capacity of a nation. It is for reasons such as these that major achievements in space contribute to national prestige [. . .]. Major successes, such as orbiting a man as the Soviets have just done, lend national prestige even though the scientific, commercial or military value of the undertaking may by ordinary standards be marginal or economically unjustified [. . .].[50]

The symbolics of space was as important for internal as for external reasons. The eventually triumphant Apollo moon landing project, prefigured by Kennedy in 1961, spread across years during which the preferred image of the United States as a place of freedom, consensus and plenty took a powerful battering in the face of events ranging from the civil rights movement to the Vietnam war. Although some were critical of the vast resources consumed by Apollo, the project provided a positive focus of attention for many Americans. Rather than being exposed as a deeply racist society at home and one that entered on lethal and unjustified colonial adventures abroad, the United States could, through Apollo, project an image to itself and others of a successful nation striving at the frontiers of human endeavour. The astronauts offered clean-cut, all-American images to balance the alternative spectacle of the 1960s counterculture. In the special message to Congress in which Kennedy announced the intention to put a man on the moon before the end of the decade he emphasized not so much the technological achievements involved in space exploration as 'the impact of this adventure on the minds of men [*sic*] everywhere who are attempting to make a determination of which road they should take.'[51] Space projects were dramatic spectacles of conspicuous consumption and sacrifice, vast sums of money going up in flames on the launch pad as a form of symbolic competition between Soviet and American chiefs. Whose wasteful expenditure could be greatest? Any material benefits were limited and wholly disproportionate to the cost.

The map tended to precede the territory in the Cold War itself, the historical background against which the space race was conducted. The Cold War was fought partly on the basis of collective fiction. A vision of the Soviet Union as a messianic force seeking global revolution and the annihilation of any opposition came gradually to displace the more plausible consideration that the rival of the United States might act as little more than another conventional player of power politics, especially under Stalin and his successors, seeking security in what it saw as its own sphere of influence.[52] Increasingly hardened and allowing little if any room for alternative perspectives, the Cold War map bore only a passing resemblance to geopolitical reality, however accurately it charted American anxieties. This was true of its initial manifestations and also later, under the post-détente regime of Ronald Reagan. There were very real differences between the two sides, of course. The Cold War was a clash between two very different systems, although not without certain points in common. Each did pose threats to the other and each had, at least in principle, a desire at some level to remap the other along its own socio-political and cultural lines. But there is an important difference between these tensions and the prospect of any immediate military threat or conspiracy. Neither was in practice about to challenge the other directly at other than peripheral points.

Local and regional complexities were erased in a bipolar mapping according to which the globe became the arena for a Manichean conflict between the forces of light and darkness. Communism was seen as single live presence spreading its ever-increasing mass of red across the map like an infection. Differences between communist or socialist regimes were largely ignored. The Soviet Union and China were presumed to be fellow global conspirators, despite differences that existed well before the formal split marked by the border clash of 1969. Only belatedly did policymakers appreciate the extent to which one might usefully be played off against the other. The break between the Soviet Union and Tito's Yugoslavia did not greatly register. Stalin also rendered the world into a binary mapping, subordinating the needs of other revolutionary movements to those of the USSR. No more than limited moral support was given to communists fighting in the Greek civil war of 1947. This did not stop the conflict being used as the launchpad of the Truman Doctrine in which the American president declared an intention to confront the Soviet Union anywhere in the world. Exaggeration was used by Truman in a conscious effort to generate domestic support for expensive projects such as Marshall aid. These were intended less to tackle the USSR than to provide orders for American business and to forestall the postwar advance of the indigenous European left. Communists who fought colonial rule in Vietnam were assumed to be stooges of the Chinese, regardless of the thousands of years of hostility between the two. Apparently just two more fallen dominoes, the neighbouring communist regimes of Vietnam and Cambodia were at war by the second half of the 1970s, while Vietnam clashed openly with China.

The lines of the Cold War proved difficult to erase once inscribed onto the map. The slightest departure was likely to be seized on by rival political factions as evidence of weakness or even betrayal. The outlines on the map had a profound effect on the territory itself. Even at its most ludicrous, the map provided the framework within which real events occurred, often with devastating consequences. It provided legitimation for imperialist interventions in various parts of the world that would otherwise be difficult to justify. The map was also used to win increased military spending by the state, a use of public resources to bolster private profit.[53] The rhetoric that descended into the grotesque caricatures of McCarthyism and the stifling of internal dissent had very potent effects both within the United States and abroad. This is a striking illustration of the way metaphor or rhetoric can create its own reality, the map its own territory.

A simplistic Cold War analysis may also have answered a need for a framework within which to understand an otherwise disorienting array of events, much like the cartographies discussed in Chapter 3. It was easier to comprehend a single master plot into which diverse events could be mapped than to attempt to understand the specific details of developments in the United States and around the world. Although it was used deliberately to justify imperialist interventions, the Cold War map was more than a cynical ideological construct. Cold warriors and ordinary citizens alike were too deeply trapped in its own grid to handle it from a distance. Many believed it to be accurate in essentials, even if it could usefully be exaggerated on occasion. If the Monroe Doctrine had drawn a line on the map that declared the American continent out of bounds to European colonialism, the period from the late nineteenth century onwards was one in which the area of United States interest widened, particularly after its involvement in the Second World War. The United States emerged in 1945 as the dominant power, both economically and militarily. The world was seen by some as a map that could be redrawn at will. It proved to be more resistant. Threats to the United States appeared to break out across the map. No longer was there a secure boundary, a line on the map beyond which the United States need not concern itself. The same seems to have applied in the postwar era to geographical boundaries (between nation-states and global spheres of influence) and those between image and reality, map and territory.

In Europe, as in many areas emerging from colonial rule, a new map had to be drawn, both of the territories themselves and of the understandings according to which their future status would be mapped. The most important line on the postwar map was drawn between East and West Germany. Like the border Keiller described in *Lake Wobegone Days*, it was supposed to be temporary, provisional, drawn merely in pencil and subject to amendment or easy erasure. Soon it was not just inked-in but marked with concrete walls, armed guards, security fences and the bodies of many of those who attempted to cross. What was once supposed to be a pencil line became not just a symbol but a very real and physical manifestation of a rapidly polarizing division into two opposed camps, only recently beginning to break down. If the strong ideas and false certainties of Cold War rhetoric were a response to underlying anxieties and uncertainties, the rigid barriers across Germany and Berlin were a defensive and anxious response to the prospect of a strategically important central European zone becoming vague and amorphous in the aftermath of war. It was more reassuring to draw a firm line, up to which each could be at least relatively sure of a ground defined as its own. The building of the Berlin Wall in 1961 averted a crisis in which the two superpowers appeared to be on collision course. It satisfied the most acute concerns of each; their subsequent commitment to reunification remained largely formal.

Cartography played its own part in the establishment of Cold War ideology. It was on the map, with its apparently neutral objectivity, that the cold warrior could most easily point out the succession of states that were supposed to become the falling dominoes of the Soviet advance. Reduced to the two dimensions of the map, neighbouring states appeared to be subject to the kind of mechanical influence suggested by the domino metaphor, or the organic infection of 'rotten apples' in a barrel. Complex social differences and other local factors were effaced. The 1939–45 war and its aftermath produced a renewed interest in cartography 'unparalleled since the time of Columbus and Magellan',[54] although perhaps rivalled since by the mapping of developments in Eastern Europe and the Soviet Union after 1989.

The use of maps to generate fear of Eastern Europe was nothing new. Frederick Rose's *Serio-Comic War Map for the Year 1877* had depicted Russia as a large octopus grasping with its tentacles at different parts of the continent.[55] A more subtle and therefore probably more effective impression of threat was created by the Mercator projection. Disproportionately enlarging the land masses of the northern hemisphere, it made the Soviet Union appear enormous and menacing. The apparent objectivity of such a projection could be used to justify a Cold War ideology that it in fact helped to create. The impact could be heightened by a selective cropping of the image to eliminate the land masses of Greenland, Alaska and the Antarctic and so reduce the area of non-communist territory included on the map. Bright colours (preferably red) could be used for the Soviet bloc to make it stand out, appearing larger than the remaining areas marked in recessive grey-tones. The finished object closely resembles an American school atlas produced in the 1950s.[56] Ideologists are nothing if not flexible, however. During congressional hearings on the proposed unification of the US armed services in 1945, an Air Force general arguing the case of Soviet threat and the central role of the Air Force in meeting it made his point by switching from the standard Mercator to a polar projection. The world took a different shape. The Soviet Union appeared to hover above the United States, a dark and threatening land mass within easy flying distance.[57]

The polar projection came to be adopted increasingly by post-Pearl Harbor American military planners as part of the new cartography that accompanied, and helped to give coherence and definition to, a growing emphasis on a more global foreign policy perspective.[58] The northern polar regions were elevated to the status of vital national interest, a move in which what started out as a propaganda map came to guide a policy of staking out the territory with early-warning radar stations. The globe itself was given renewed prominence by those for whom American interests spread across the world regardless of either continental borders or the confines of one flat map or another. Globes had their own limitations, however, not the least of which was their cumbersome shape and the expense of producing anything large enough to provide a high level of detail. The wartime maps produced by Richard Eades Harrison went some way to replicating the global view on paper. Widely reproduced in *Fortune* magazine, Harrison's maps were perspective views of sections of the globe as if taken from the stratosphere before the days of satellite photography. They were also sold to the United States armed forces and used by Air Force bomber and reconnaissance crews.[59]

The polar projection could be used by the other side to produce an image of the Soviet Union encircled by enemies. Rather than looming across the top of a Mercator projection it becomes vulnerably enclosed, an effect emphasized by adding the concentric circles of surrounding lines of latitude. The use of colour again completes the effect. This time the Soviet Union is depicted in light colour surrounded by the darker or brighter shading of Western allies and other countries.[60] Soviet cartographers also deliberately falsified the location of certain strategic sites, a tactic made less effective by developments in satellite photographic technology.[61] Such misleading practices may also be found at a more mundane level. It is not unknown for commercial map-makers to include small fictional additions on their works as a way of proving any subsequent cases of copyright theft.[62]

The end of the Cold War has also been celebrated on the map. Within weeks of unification the German government was circulating and promoting a map of the new republic, symbolically to reinforce the reality of what remained a new and raw construct. It also helped to create an impression of restored unity and wholeness in a new state that was wrought by divisions, especially as some in the east gained little more from unification than immediate unemployment. The new German map did not show much of the territory of its neighbours. In this way it contrasted with the maps drawn up by those for whom the new geographical entity created a lapse into older ways of thinking about the European order. Here the new Germany was shown alongside other countries as a large and potentially threatening presence. Cold War or none, maps remained powerful tools of political and ideological debate.

I want to return now to the question of the structure of *The Right Stuff* and the kind of opposition it suggests. This does initially seem to accord with the temporal schema set out by Boorstin or Baudrillard. A movement is traced from an earlier more authentic mode of experience to one in which mediation has taken over. A similar emphasis is found in Walter McDougall's history of the space race. Unlike Eisenhower, he recognizes the impact of matters of symbolism, status and prestige. He accepts that they can and did in this case come to override factors such as actual military strength or technological ability. Yet McDougall's version remains close to the spirit of Boorstin, who is cited approvingly. For McDougall: 'The concomitant arrival of Sputnik and the Third World generalized the problem of the American image [. . .].'[63] The United States had to present the right image to the emerging nations: 'This meant the extension to foreign policy of a decadence in the United States that was the subject of Boorstin's book. [. . .] The brief Kennedy years were those in which American space policy fell captive to the image makers.'[64]

The importance of the image, or the inability fully to disentangle it from any separate reality, is again seen as a merely localized phenomenon, the extension of a decadence or corruption rather than a fact in its own right. An alternative reading is possible. The key issue is the way *The Right Stuff* signifies the supposed authenticity of the test-pilot experience. The world of the test-pilot is associated with that of western frontier mythology, a mythology that is itself rooted in questionable notions of authenticity and immediacy. It is also a mythology that recurs frequently within the space imaginary to which it seems to be opposed in *The Right Stuff*.

6 Mapping the Frontier

The frontier signifies the decisive exclusion of all that is not culturally familiar: and it excludes it even when it incorporates it. [. . .] Savages are, by definition, what are found beyond the pale of civilization: there is no question of letting them dispense with this boundary or turn their backs on it, for this would undermine the imperial logic of opposites on which the frontier myth rests.

Paul Carter[1]

The desert base from which Chuck Yeager flies to break the sound barrier in *The Right Stuff* is depicted as lying on the outer edges of civilization. The milieu is that of the western frontier settlement. The pilots drink, ritually, in Pancho's bar, a flyblown inn owned and run by a tough frontier character with a sharp tongue. The surrounding landscape is of Joshua trees, scrub and barren desert. In the film it is all picked out in elegiac, autumnal colours, suggestive of the end of the frontier era; browns, red-browns, heavy shadows and silhouettes dominate, in contrast to the bright, clinical, technological look of the astronaut sequences. An explicit link is made between the art of the test pilot and that of the western horseman. Yeager himself is a daring rider and his high-performance aircraft is likened to a wild horse. The test-pilot rides bareback, we are told, without the mediating presence of any saddle. In the film we see Yeager riding out to a corner of the airfield where the X-1 jet is being fuelled, a brooding presence with an aura that makes the air shimmer. Yeager's horse is spooked, but that we are to equate it with its metallic counterpart seems clear. The X-1 is the next bronco that will have to be broken or will kill the man who tries.

The frontier, a concept deeply rooted in American mythology, has often been offered as the authentic American experience. The frontier line is seen as marking the point at which culture and nature intersect. For Frederick Jackson Turner, in his influential essay 'The Significance of the Frontier in American History' (1893), it is 'the meeting point between savagery and civilization'.[2] The American frontier is distinguished from those between different states in Europe in that it 'lies at the hither edge of free land',[3] rather than passing through highly populated regions. At the ever-advancing American frontier we find 'the complex European life sharply precipitated by the wilderness into the simplicity of primitive conditions.'[4]

However different the American frontier might have been from those found in Europe, it did not entail any point of contact with a primordial ground, an existential encounter with something immediate and unmapped. Such an illusion, established in many works of both elite and popular culture, can only be maintained by assimilating the pre-existing Native American cultures to nature or 'savagery', notions that were functional for those seeking to justify further encroachments into native lands, forestalling the idea that what was involved was a form of cultural genocide rather than a mastering of empty wilds. Much of the north American landscape occupied by the British, French and Dutch from the seventeenth century may have appeared devoid of native occupants. This was not because it was by nature an unpeopled wilderness. Huge numbers of indigenous peoples had already been wiped out, primarily by disease, as a result of earlier European adventures: 'The American land was more like a widow than a virgin.'[5] The popular conception of the frontier also obscures close connections between the rural landscape and the central metropolis. Rather than being in some way primal or originary, events on the American frontier were largely structured by the dynamics of the metropolis, whether in the shape of the European colonial power or the cities of the eastern seaboard. Regional centres such as Chicago organized the activity of a vast western hinterland, calling into being rather than growing out of great domains of frontier life.[6]

The frontier is usually seen as a line continuously pushed forward by the heroic pioneer. The line exists only on the cultural map, but maps both cartographic and conceptual played an important part in the process of colonial domination. The frontier line as it appears on the map is akin to the line drawn between the map itself and the territory: an arbitrary construct across which are organized an array of cultural meanings, including articulations of the real and the unreal, the authentic and the inauthentic. The presence of the indigenous peoples on the land as any kind of active subjects is usually denied in frontier discourse, cartographic or otherwise. Large areas of the map were often left blank, as if empty and available for settlement. The Native American presence was shown on some maps, although it could be subjected to a process of removal similar to that enacted on the ground. Two large wigwams are planted firmly on the landscape of Jean Rotz's 'Map of North America and the West Indies' (1542), based on the findings of the first voyage of Jacques Cartier to the New World.[7] Along with trees and figures carrying bows and arrows, they give the impression of a land occupied rather than empty. A number of native figures are drawn onto the territory in the 'Harleian Mappamondo' (*c.* 1544) attributed to Pierre Desceliers and showing the findings of Cartier's first and second voyages.[8] The word 'Canada' appears three times, a local term for a collection of houses perhaps still reflecting the presence of indigenous settlements rather than being translated as later into the language of colonial appropriation. A number of native figures engaged in various activities populate a map of 'La Nuova Francia' based on Verrazzano's voyage of 1524 and reproduced in Giovanni Baptista Ramusio's collection *Delle Navigationi et Viaggi* (1556).[9] Shelters comprised of upright posts and thatched roofs are also depicted, again suggesting more than a passing occupation of the territory. Both this and the previous example also show the figures of European explorers whose followers were to displace the natives, a process that can also be traced in subsequent maps.

The only native figures in the 1612 map of Virginia by John Smith, eventual leader of the early British colony of Jamestown, appear on the margins.[10] A vignette of the council of chief Wahunsonacock (known to the English as Powhatan) is presented in a box in the upper left corner. A larger figure appears in the cartouche opposite, his feet just about on the edge of the territory although his body occupies no more than decorative space. On John Speed's map of the Americas in *A Prospect of the Most Famous Parts of the World* (1627) the indigenous population is removed entirely from the territory, limited to a series of illustrations in the borders on either side.[11] All that remain on Augustine Herrman's map of 'Virginia and Maryland' (1673) are two figures in the cartouche, standing beneath the title on a ledge as narrow as the ground that would be left to their counterparts on the territory. There is little suggestion of the sovereignty over the land often symbolized by the figures of European explorers or monarchs in such decorative positions. Many later mappings erased the native presence altogether.

The maps of many explorers could not have been drawn without help from the indigenous peoples.[12] The master map of the West kept by William Clark in his office at St Louis, which was to prove one of the most important in the nineteenth-century mapping of that previously hazy part of the continent, was based largely on information from Native Americans whose maps sketched on hides or on the ground were more accurate and detailed than those from other sources.[13] Such contributions were seldom acknowledged in the drive to impose a Western mapping upon both the territory itself and any pre-existing native cartography. Colonial maps often included names derived from the languages of the indigenous peoples, whether or not they were used to label the same features by those peoples themselves. Smith's map of Virginia inscribed about 200 names. Some were English but many were of Native American origin, including the name of the Patawomech (Potomac) river which survived onto the modern map of the United States.[14] There were pressures to change the Native American names, however, as can be seen from Smith's map of New England produced two years later. The earlier names Penobscot and Sagadahock have here been replaced by the Scottish Aborden and Leth, according to the demands of the then fifteen-year-old Prince Charles of Scotland.[15] Smith's map of Virginia was itself superseded by Herrman's 'Virginia and Maryland' in which many of the persisting Native American place-names were supplanted by English. The topographical detail did not change, but such was the importance of the process of naming in the realization of the territory as colonial property that Herrman's map became the new prototype chart of the area, a status it maintained for more than three-quarters of a century.[16]

That which lies across the frontier line is in some cases experienced as an unreal 'other' that guarantees the reality of civilization. Alternatively it may be the so-called 'primitive', on the far side of the frontier, that is accorded the status of the real or the authentic. A desire for access to such a reality is a recurrent feature of the American imaginary. The first settlers were seen as escaping the stifling artificiality of European cultures. The locus of the real shifted westwards as substantial towns grew on the eastern seaboard and the landscape was tamed, its capacity to provide authentic experience seemingly diminished. Eventually the frontier between real and unreal moved out to the west coast and the far western plains and mountains. Today, perhaps, it has doubled back as the west coast (California in particular) is deemed the ultimate landscape of artifice. A pivoting between alternative articulations of the real and the unreal, the authentic and the inauthentic, thus occurs across the frontier boundary. That such constructions can function in either of two opposite directions underlines the arbitrariness of their nature. It does not reduce the potency of their appeal.

The frontier experience generated a need for maps, both physical and conceptual. Many owed more to fantasy and invention than to the character of the territory itself, as we have seen, but this did not make them any the less influential. The grids that were imposed aided both the sale of the territory and efforts to make it signify. A particular problem in America was the vagueness of the land, not just because of its size but because of the undefined nature of its borders. For many decades the geography of the interior was unknown to European settlers and the borders shifting and uncertain as new territories were acquired or fought over. New maps had constantly to be prepared to define the extent of the post-revolutionary nation and its increasing number of member states. The resulting vagueness of American borders was for Boorstin a potent source not only of American myths and illusions but also of the nation's optimism and energy; it 'would long profit from being born without ever being conceived.'[17] Like Israel in the twentieth century, the United States came into *de facto* existence as a state before its boundaries were clearly defined. But it was far from being born unconceived. What was to become America, both north and south, had been *pre*-conceived long before the landing of Columbus or the Mayflower, victory in the war of the revolution or the ratification of the Constitution. The existence of uncharted territories in the oceans of the west was well established in mythology. From at least the ancient Greeks onwards there had been tell of the existence of a western paradise, from Homer's Elysian Fields and Hesiod's Hesperides to Plato's isle of Atlantis and the legendary lands of Avalon and

Lyonesse.[18] When European exiles headed west in search of a better world they partook of a venerable tradition.

In none of these cases, real or imaginary, was there any unmediated contact with the New World. Accounts and understandings of the Americas were no less a function of the particular perspectives of their conquerors than the description of Atlantis given by Plato was rooted in his own metaphysics. America was thus seen in terms of a series of Renaissance conventions. Some commentators have argued that the Europeans were aided in their combat with the natives by a superior access to reality. Samuel Purchas, writing in the early seventeenth century, argued that the literacy of the Europeans marked their superiority over the natives just as the power of speech distinguished 'man' in general from the beast. Literacy was said to have provided the basis of history, an ability to escape from an existence rooted only in the present.[19] Tzvetan Todorov suggests that the possession of writing and an improvisational relationship to signs that could be manipulated in new ways gave the Spanish invaders a crucial advantage over Aztec and other native cultures. Aztec communication, he suggests, was essentially verbal and ritual, obeying an orderly cyclical routine. All events were believed to be foreseeable, even if the prediction was often filled in retrospectively as a way of fitting the new and unexpected into the existing cultural map.[20]

Interpretations of the new in the terms of the familiar are hardly unusual. According to Todorov, however, it was an inability to comprehend what was really involved – the arrival of greedy outsiders prepared to extinguish an entire civilization – that rendered the Aztecs impotent and indecisive in the face of the enemy. Cortés, in contrast, was aware of the importance of manipulating signs and appearances to increase the confusion of the Aztecs, as is clear in places from his own writing. He knew the value of spectacular actions such as the construction of a giant catapult during the assault on the city of Tenochtitlan: 'Even if it were to have no other effect, which indeed it had not, the terror it caused was so great that we thought the enemy might surrender.'[21] It was by his mastery of signs, Todorov concludes, that Cortés ensured control over the ancient Mexican empire.

Any tendency the Aztecs might have had to interpret the new and strange in terms of the old and familiar, or to struggle to achieve any satisfactory interpretation at all, does not seem to be a suitable basis on which to distinguish them from the European invaders. The Aztecs may have misunderstood the identity and motives of Cortés and his fellow conquistadores, but the Europeans were hardly guilty of incisively perceiving the realities of the New World. If anything their success in military-colonial terms might have been more closely associated with their inability to see the radically different in any terms other that those dictated by their own maps. The letter written by Columbus in 1493 that contained the first reliable report of the actual existence of the New World (whatever its author's remaining doubts) was singularly lacking in specific description of place, as was the journal he kept. What is found instead is a bland repetition of formula. Every island is 'very green and fertile', each successor 'much more so' or 'the loveliest thing in the world'.[22] What Columbus and other writers described was largely preconditioned by the model of the world they carried with them. Such words provide little upon which we can rely for accurate information about the peoples or places visited by the likes of Columbus. As Stephen Greenblatt suggests, they are a record primarily of the customary representational forms of the time.[23]

Empirical observations served principally to confirm what was already mapped. Columbus knew in advance, from the maps and globes to which he refers in his own accounts, where he would arrive. Or at least he thought he did, accepting a mapping by Pierre d'Ailly in which the size of Asia was exaggerated and the width of the ocean was grossly underestimated. A view of the world close to that of Columbus was reproduced on Martin Behaim's globe of 1492. The distance from the Canary islands to Japan is calculated at 2400 miles rather than the actual distance of 10,600. If it had not been for the existence of such erroneous charts Columbus might never have been induced, or permitted, to set out in the first place. Once on land he continued to apply a mapping that had been determined in advance. Gold and rich spices would be found in abundance, he had predicted (or claimed) when seeking backing for the venture. Everywhere he looked, accordingly, he found signs of both, although his hopes were often ill-founded.

The contact between Europe and the Americas that resulted from the voyage of Columbus marked a revolutionary break in the history of the West. It involved a confrontation with radical otherness more acute than any that had gone before. Contact with Africa and the East had been gradual and incremental, a process that spread over many years. Yet the American experience was made to confirm existing assumptions as much as to provide an opening to the new. Columbus himself refused to be shaken from his preconceptions. He maintained, for example, that Cuba was not an island but a peninsula of Asia. When the natives told him otherwise it was their reliability that was questioned rather than the 'truth' already possessed by the explorer. More than the reality in front of him, the authoritative voice of writers such as Marco Polo and John Mandeville guided his perceptions. The works of these writers, in turn, were embedded in layers of mythology and fabrication. Polo's account of his *Travels*, co-written by a writer of fictional romance, is of uncertain veracity. Whether Mandeville even existed as a historical rather than imaginary figure remains unclear. In both Polo and Mandeville authenticity is asserted repeatedly through the claim of the observer to have been there and to have seen the marvellous for himself, the kind of travel experience Daniel Boorstin might be expected to celebrate as opposed to the shallow and mediated experience of the modern tourist. This is a rhetorical device, however, wielded in an attempt to persuade the reader to accept that which might be the product of imagination or exaggeration, rather than any guarantor of truth.

Even where the authors of the travel narrative were actually present at the scenes described, and writing for themselves without outside assistance, they were inevitably involved in a process as much of calling into being as representing an external world. However much their works might have led to attempts to redraw the map of the world, they could not be expected to escape the perspectives of their own cultural cartographies. They might be able to describe with relative indifference some social practices and sexual habits radically different from those with which they were familiar at home, but there were limits to what could be seen or how it could be fitted into existing grids. When Columbus interacted with the natives of the Caribbean he seemed incapable of appreciating the possibility of a real diversity of language and the different social reality that might imply. The 'other' was either deemed to have no language at all worth the name (and in some cases to be taken back to Spain 'to learn to speak') or the language that did exist was assumed to be intelligible to the colonist. Hence the ability of Columbus to engage in absurd 'dialogues', interpreting whatever was said by the natives to further confirm his existing assumptions on subjects such as the whereabouts of the Great Khan from the pages of Marco Polo.[24] He was free to pick and choose, conveniently, just what could or could not be understood, at times shifting register in the space of consecutive sentences, as in this report of exchanges with a chief during the first voyage: 'He and his councillors were extremely sorry that they could not understand me, nor I them. Nevertheless I understood him to say that if there was anything I wanted, the whole island was at my disposal.'[25] Other peoples were either essentially the same as the Europeans, and so assimilable to Christianity, or different in terms only of inferiority. Columbus gradually moved from the former to the latter, from a delight in the friendliness of natives who seemed immediately open to conversion, to their enslavement, torture and death.

On his third voyage, in the presence of large volumes of fresh water flowing from what we now know as the Orinoco river, Columbus was convinced that he was close to the biblical Paradise from which the four great rivers of the world were said to flow. He sketched a bizarre biblico-mapping suggesting that the earth was not spherical but pear-shaped, 'or that it is like a round ball, on part of which is something like a woman's nipple.'[26] If the source of the river was not Paradise, he wrote, 'the marvel is still greater. For I do not believe there is so great and deep a river anywhere in the world.'[27] The alternative would be that the river came from 'a vast land lying to the south',[28] but Columbus was convinced that it was from Paradise: a safer, more familiar conclusion than the recognition of a whole new continent.[29] Such mappings might have to be modified in the future, as closer engagements with the New World rendered elements of the mythology unsustainable. Often it was the reality on the ground that was changed. Some way had to be found to generate wealth from the islands explored by Columbus to justify the costs and reputations staked on the enterprise. When they failed to provide an easy supply of gold, spices and the other anticipated fruits of paradise the alternative option was to settle and colonize. European forms of land use and tenure were imported wholesale, adapting the territory to fit the profit map.[30]

The first word of the New World for many Europeans came in Latin translations of Peter Martyr D'Anghera's journal *Decades*, in which the reports of explorers such as Columbus and Vespucci had already been translated into a Renaissance language that made them evoke the garden landscapes of the Golden Age.[31] The Spanish went on to describe their bloody engagements with the native populations in the terms of chivalric romance, while to the English Puritans, as we have seen, their new territory seemed to embody religious strictures about the dangers of unruly wilderness. For Columbus and his contemporaries the situation can be seen in terms similar to those used to discuss reassertions of order and control amid the experience of the modern or postmodern. Their narratives came at a time when the dissolution of elements of the existing order created disorientation and unrest. Onto the New World, previously the subject of endless speculation but now discovered as a firm reality, could be mapped a territory more reassuring in aspect and in its apparent confirmation of existing dominant assumptions.

Myths, illusions and inventions went on to play a central role in the creation of post-colonial American identity. Emerging American linguistic forms were dominated by a colourful and exaggerated form of tall-talk that 'blurred the edges of fact and fiction'.[32] As settlers and speculators sought to promote their own towns, booster-talk, a language of optimistic overstatement and anticipation, found its way onto the map of the country in the names of places such as the (short-lived) Wealthy City, Kansas, and a glut of settlements named after Old World settlements including Oxford, Cambridge, Rome and Carthage – or even Paradise, Montana. For those who did the christening the hope was that the image could generate the reality, that the territory could flow from the map in a self-fulfilling prophecy of prosperity and high culture. Many places were in fact named before they even existed, brought into being with the map onto which their titles were inscribed, places like Keillor's Lake Wobegone, reached by its founders only because they misread the map and built speculatively: 'not so much a city as a trance, a whimsy built upon a swamp, a steeple waiting for its church, a naked man in a fine silk hat.'[33] Chicago was first mapped out into city lots as a result of a purely speculative boom in the 1830s following the start of work on a canal to link the watersheds of the Great Lakes and the Mississippi – another attempt to create the great transcontinental waterway of American mythology.[34]

Against the realities of state and sectional fragmentation and rivalry, American unity was forged largely in the symbolic and mythical dimensions. Those for whom the New World offered the prospect of a non-mythological existence, escaping from such inheritances in Europe to a land bathed in a clear light of rationality, were simply the victims of one particularly potent myth. Narratives of origin are found in most cultural groups, not least the Native American inhabitants. Such accounts help to define what is peculiar to any given group. The greater the threat of confusion or fragmentation the more regimented or prescriptive such narratives may have to be if they are successfully to articulate unifying principles. Narratives of national or cultural identity may take a variety of forms, oral or written. In the United States, growing up in the age of the printing press, literary forms were to play a dominant role. Conscious efforts were made to fabricate a national myth. As Richard Slotkin suggests, poets brought up on a classical diet of Homeric epic sought to construct their own American equivalents. Works such as Joel Barlow's *Columbiad* (1787) and Timothy Dwight's *Greenfield Hill* (1794) began a process continued in later years by writers including Mark Twain, Herman Melville and Walt Whitman. Literary epics had only a limited power to make connections with the lives of ordinary American settlers, to gain the impact and resonance required if the status of myth was to be achieved. Other narrative forms such as religious sermons and accounts of interactions with Native Americans proved more effective, reaching a wider audience and tackling more directly the pressing issues of the day.[35]

For a viable notion of the American to be constructed, some kind of negotiation had to be achieved between Old and New Worlds. At first rigid Old World conventions were imposed, in most cases with little or no regard for the specific realities of the new. This process found its most extreme form in the sermons and writings of the Puritans in New England, where the experience of the new continent was reduced to an enactment of religious dogma, a simple and dramatic confrontation of opposites: Christian and 'Indian'/heathen. Where Native Americans proved helpful to the settlers it was because they were agents used by God to bring succour to the faithful. Otherwise they were servants of Satan. In neither case were they allowed an existence of their own as members of different cultures.

In the first years of settlement this rigid mythological construct may have been functional, creating an opaque screen between the settlers and what might otherwise have been an overwhelming experience. In time, as Slotkin suggests, it proved a barrier to success. According to its terms those who in any way threatened to cross the divide and come too close to Native Americans or their ways were to be condemned, in some cases to death. Accusations of witchcraft were often made, as we saw in Chapter 3, in an attempt to account for the actions of those who failed to fit into the dominant grid. As articulated by preachers such as Increase and Cotton Mather it was an inflexible schema that allowed little or no leeway or scope for innovation. It was also hostile to the economic and geographical expansion inherent in the capitalist aspect of the colonial project. Where contacts were made with native cultures Puritan ideology sought to prevent any real process of communication. When settlers were captured and obliged to live in Native American societies the resulting experiences were not usually offered as insights into other valid modes of existence. Instead, they were fitted neatly into religious convention. The Puritan settlements were thus kept from learning many of the lessons necessary if they were to thrive in their new environment. Only a limited succour could be gained from the clearance of small enclaves in what was perceived to be an alien wilderness in terms of both geography and the soul. Imposed on the land with almost no rooted connection to the territory, many settlements failed. To thrive they needed to adapt more to the territory, even if the extent to which this could be achieved was limited and a more abstract grid was later to be reimposed once the colonists had succeeded in asserting their military superiority and shifting the contest onto a terrain of their own.

A new mythology had to be developed if such a change was to occur. Slotkin suggests that accounts of wars with the natives such as that written by Benjamin Church in 1716 forced a recognition of the need to value native ways when fighting on their ground. Church is seen as a pivotal figure in the movement from one mythology to another, able to impose his own order onto events in the wilderness because he had gained a greater understanding of the environment. Like James Fenimore Cooper's Leatherstocking, the kind of mythical figure who was to follow stealthily in his footsteps, Church was able to pick what seemed most appropriate from Puritan or Native American cultures: 'The Puritan townsman, in contrast, believes the wilderness is a chaos because he does not understand the laws of survival in the forest.'[36] The new mythology, one of lasting resonance, was that of the frontiersman: essentially white but equipped with a respect for and facility in certain strategically valuable native ways, and to gain its widest recognition in the legendary figure of Daniel Boone.

The mythology of the frontiersman or backwoodsman may have been based upon a greater engagement with the realities of the American landscape than its predecessors, but it remained rooted also in fiction, and it was this dimension that provided its resonance. Daniel Boone, probably not an outstanding historical figure, was created to a large extent in the literary imagination. He owed his fame primarily to John Filson's *The Discovery, Settlement and Present State of Kentucke* (1784), to which was appended a narrative of Boone's adventures. Despite Filson's claim to have published the work 'solely to inform the world of the happy climate and plentiful soil of this favoured region' and not at all 'from lucrative motives', it was little more than an elaborate real-estate promotion brochure designed to sell farm lands to easterners and Europeans.[37] A Pennsylvania schoolteacher, Filson travelled to Kentucky in 1782 and began surveying to produce the map of the territory that accompanies his text. His map is accurate, he claims, unlike its predecessors, and it plays a key role in underpinning the work. In the text, Filson refers frequently to details as shown on the map. As Slotkin suggests:

> The structural plan and argument of *Kentucke* are modelled on those of the Puritan narratives and histories. But where the traditional sermon form begins with a biblical text, Filson takes the map of Kentucky for his text. His plan is to develop the meaning inherent in the land in much the same way that the Puritan sermon exfoliates the meaning in the biblical passage. The map itself is watermarked with a plowshare

> and the words 'Work and be Rich'. By holding the map up to the light, the alert reader can thus see behind the pattern of the map the substance of Filson's doctrine.[38]

Book and map were published in 1784, the year that also saw the start of the great rectangular survey of the west. In its way Filson's exercise in cultural cartography was to prove almost as influential as the physical survey of the American territory.

Filson's Boone undergoes a series of initiations into the ways of the wilderness, learning lessons that enable him to benefit his society on return to civilization. Whatever the merits of his map of Kentucky, Filson's portrait of Boone provided a grid through which it is possible to map one chapter in the emergence of the national American consciousness. Its terms were increasingly to be articulated in the pages of fiction, the development of the myth coinciding with a growth in the market for such narratives. In the Filson version the man of action is equipped with a philosophical awareness and an ability to contemplate both the majesty and terror of the wilderness environment. A range of variants followed according to different points of view and rival notions of what it was, or should be, to be American. Radicals presented the Boone character as the perfect example of the natural man of reason. For conservatives he was symptomatic of a degenerate society regressing from the pitch of civilization achieved in Europe.[39] To maintain credibility in America the debate in the colonial and subsequent periods had to conform in more detail to existing and ever new sources of knowledge about the west. This knowledge remained strongly conventionalized, however. As sectional differences became more pronounced, the Boone character was again brought into play to illustrate differing visions of the national culture. For writers in the west, Boone was a man of action, a heroic wilderness hunter and explorer stripped of the philosophical baggage from Filson that had so much impressed his European interpreters. This characterization reached its peak in the largely fictional creation of Davy Crockett as a vernacular frontier hero, a process closely linked with the development of Jacksonian mass politics and the entry of American politics into the realm of popular spectacle.[40] In the eastern states, now a more stable civilization situated midway both physically and culturally between Europe and the west, Boone might be either the degenerate embodiment of a dangerous Jacobinism or be cleaned up to suit the more refined eastern palate.[41] The real historical figure of Daniel Boone became caught up in the process of fictionalization. Having been shaped by Filson into a mould little resembling the original, Boone ended up modelling his own public statements on Filson's characterization. A similar process was undergone by another western hero, Buffalo Bill (William Cody), who ended up playing his fictionalized self on the stage and then wearing the elaborate theatrical costumes offstage: it reached the point 'where no one – least of all the man himself – could say where the actual left off and the dime novel fiction began.'[42]

Christopher Columbus, along with George Washington the most important and enduring American hero, was also elevated to his position of celebrity through a gradual process of fictionalization and through the inscription of his name onto the map of both the continent and the national psyche. Columbus was almost certainly not the first European to set foot on the New World, although his enterprise did provide the model for the colonial process that was to follow. This could not be said of the earlier landfall of the Vikings, some five hundred years before Columbus, or of the Bristol fishermen believed to have ventured as far as the Newfoundland coast in search of cod. John Cabot may also have rivalled Columbus in his own era, but his explorations went largely unrecorded at the time and his fleet was eventually lost at sea. What assured Columbus his heroic status was not only the fact that he survived and began the Spanish conquest but that he kept a journal, a daily record of his travels that enabled them to pass more easily into history.[43]

The content of Columbus's journal was slight, largely a blend of preconceived illusion, unreliable reportage and fabrications designed to justify his mission and to make up for the initial failure to find any significant amounts of the gold he had promised. The only version available is not the original but an abstract made by Bartholomé de Las Casas, which often summarizes rather than quoting directly from the text. But the journal provided the outline of a map that was over the years to be filled in greater detail by others. His son Hernando, who accompanied Columbus on his fourth and final voyage, made one of the greatest contributions with a full-length biography that offered a flattering portrait and became the principal source for future Columbus students seeking access to the detailed colour and flavour of the enterprise, although here again doubt has been cast on the authenticity of parts of the work.[44] Peter Martyr's *Decades*, published in numerous editions from 1504, devoted much of its space to an account of the explorer and his heroic achievements. His version was lifted, sometimes verbatim, by successors such as Sebastian Munster, whose popular *Cosmographia* went through 35 editions between 1544 and 1576.[45] Others paid tribute to Columbus in epic poems, often of dubious literary and historical merit.

Translations of these works made the Columbian legacy available to the first settlers in the northern part of the continent. It was less the real Columbus than the heroic and largely fictional image that provided a mythic point of reference for those heading into unknown territory and a model of fortitude and perseverance when times became hard. Columbus was also seen as the quintessential heroic American individual standing alone against outmoded tradition, authority, superstition and prejudice in his determination to cross the Atlantic, despite the fact that his own mappings were inaccurate and often steeped in biblical fantasy.[46] His name soon began to appear on the map of North America, while early in the nineteenth century the old debate about the name of the country was reopened with calls for it to be changed to the United States of Columbia, or just plain Columbia. Those who supported such a change were forestalled in 1819 when the former colony of Nueva Granada, one of the first in South America to become independent of Spain, renamed itself Colombia Grande, later to become present-day Colombia.[47] Eventually Columbus became no less than the idealized personification of America in the allegorical figure of Columbia, an appropriate choice for the nation that had so faithfully followed through the logic of the colonial process he began. The figure of Columbus could also be inverted and made to stand for something else. To those who opposed the festivities marking the 500th anniversary celebrations in 1992 he symbolized an ongoing process of imperial domination.

A survey in 1988 found some 65 geopolitical entities on the map of the United States bearing the name Columbus or Columbia, not to mention uncounted numbers of roads, parks, squares, buildings and other features.[48] The explorer's name has also found its way onto the maps of many other countries, rivalled only by that of Queen Victoria in geographical number and spread. On the map of America the only real competition comes from the name of Washington. While Columbus became a national hero in a slow process of gradual accumulation over many years, George Washington was elevated to the pantheon from the realm of earthly political controversy with indecent haste, achieving mythological status within a few years of his death. The process was begun by Mason Weems's popular biography of 1809, which included large chunks of fictive invention. Previous works on the first president had been worthy, dull and weighty to the average reader. Weems set out to create a human figure with whom the growing mass audience could identify. To symbolize the virtue of the Union at a time when its foundation was uncertain in the face of British opposition and sectional differences, its hero had to be beyond reproach. Weems obliged with a portrait of unalloyed virtue.[49] His Washington was another figure bearing the imprint of Filson's Boone. Like Filson himself, Washington was initiated into the wilderness as a surveyor before going on to play his part in the mapping of American identity. Other cultural heroes were subjected to a similar process of *a posteriori* fabrication. Patrick Henry's famous revolutionary declaration 'Give me liberty or give me death' was invented later by his promoter, William Wirt; James Otis's celebrated 'Taxation without representation is a tyranny' was probably of equally dubious authenticity.[50] The fourth of July was not, as Boorstin recalls, the day on which the Declaration of Independence was actually signed, and the famous Liberty Bell – 'the Holy Grail of the American Revolution'[51] – sounded only in the pages of fiction or myth. Even the stars and stripes, that most powerful icon of modern American patriotism, was a relatively late construction, as was the national anthem. From the very start the idea of the United States as a real unified entity, and the terms in which it was celebrated, were caught up in a process of fiction and simulation. Sensitivity about the sanctity of symbols such as the flag is testament not to the strength but the fragility of the dominant American cultural cartography.

The frontier line on the map also played a part in the crystallization of the nation. For Turner the experience of the frontier was an important consolidating agent. His emphasis is primarily military: 'The Indian was a common danger, demanding united action.'[52] But the frontier was probably more important in terms of cultural and ideological unification, providing in the form of the much-maligned 'savage' the alien other against which American civilization could more easily be defined and articulated. That the frontier experience continues to have so great a resonance in the American imaginary would tend to confirm such a suspicion. The pure western is now something of a rarity in the cinema, although it has been maintained in a variety of other forms. The frontier, and the array of meanings articulated around the line on the map, has been relocated both inwards and further outside. Inwardly, the western has been transposed into the urban-western where the 'wilderness' to be tamed becomes that of the bleak landscape of the inner city. Frontier imagery is also found in media accounts of real inner city disturbances. The effect is to expel to the periphery, to what is often presented as the pathological actions of a few outside anti-social 'savages', that which is really a symptom of malaise directly applicable to the main social body – racism, social deprivation, the impact of public spending cuts, and so on. This process began even before the official announcement that the frontier was closed when it disappeared from the population density map of the United States in the census of 1890. A frontier mapping was applied when violence was used against a series of strikes that came to a head in 1877. Native American and striker alike were defined as a threat to society, alien 'savages' against whom the use of force would be legitimate. New technologies of virtual reality and 'cyberspace' have also been described in terms of a new inward frontier, between the computer matrix and the human brain. Frontier rhetoric can also be found across America in architectural form, with mock western town frontages gracing buildings as diverse as glittering Las Vegas casinos and the facades of the smallest of small-town shops. These are simulacra of structures that were already facades in the original, one-dimensional fronts that signified booster aspirations towards greater things. From the side or the rear they could be seen to be little more than cardboard cut-out fakes, hardly less solid than the western movie sets on the studio backlot that were to ensure their place in the modern imaginary.

In an outward direction, the frontier is projected into space, the 'final frontier' of *Star Trek* or the high frontier that became of such concern during the space race. Cinema again abounds with examples. The spate of American science fiction films in the 1950s was 'an extreme extension of established frontier myth' in which were rehearsed a variety of popular uncertainties and perceived threats.[53] Boorstin seems to miss the point in his reading of the relationship between the space project and the frontier experience. The original western frontier alone remains authentic, he declares. 'The great deeds of our time are now accomplished on *unintelligible frontiers*. Even the most dramatic, best-publicized adventures into space are on the edge of our comprehension. [. . .] Fantastic possibilities engage our imagination without taxing our understanding.'[54] Of course few of us can claim to understand the technological detail, but this is far from denying the space project a cultural meaning, at least in part as a contribution to the maintenance of frontier mythology. In space an attempt might be made to chart a new frontier landscape of the authentically real that has been lost on earth. The connection is made felicitously in a Texas poster on sale at the NASA headquarters in Houston. An astronaut floats high above the earth in a familiar image but one that has undergone subtle, wild-westernizing change: the cable by which he is attached has been replaced by a rope, the box strapped to his chest is in studded leather rather than silver, and a denim pouch is strapped to one leg; on his hands are leather gauntlets, his feet are packed into cowboy boots and, although a space helmet is on his head it is a stetson that is grasped in his hand. 'Texas', the poster declares, 'still in the frontier business.'

When John Glenn became the first American to go into orbit he was welcomed by Lyndon Johnson as the new Columbus, a reappropriation of a pioneer mythology that had cheekily been stolen by the Soviets when Gagarin was greeted after his earlier flight as the 'Columbus of the Cosmos'.[55] For Thomas Paine, then head of NASA, the experience of the moon landing (and the subsequent settlement of the moon that he predicted) was to be likened to that of the founding fathers. 'As with the American experience of 1776, founding a new society in a demanding environment will sweep aside old world dogmas, prejudices, outworn traditions, and oppressive ideologies,' he enthused.

'A modern frontier brotherhood will develop as the new society works together to tame its undeveloped planet for posterity. Advances in extraterrestrial societies will surely be reflected back on earth.'[56] But if anything it was earth-bound issues that were reflected in the exploration of space, while Paine's words remained very much a part of the kind of dogma, prejudice and ideology from which he sought to distance himself. The historical resonances intended in Kennedy's promise of a 'New Frontier' could hardly go amiss, either, nor fail to chime with the promised exploration of that infinite frontier in space. Christa McAuliffe, the ill-fated recipient of the first citizen flight in the space shuttle *Challenger*, had in her application compared the experience with that of the early women pioneers. The impact of the *Challenger* disaster on the collective psyche only further demonstrated the continuing currency of frontier discourse.

A conceptual link can be made between the space programme and the early voyages to the Americas. At the time of Christopher Columbus and Amerigo Vespucci the accepted view of the world was that of the *Orbis Terrarum*, the 'circle of lands'. According to religious dogma the *Orbis*, consisting of the linked land masses of Europe, Africa and Asia, was the portion of the globe allotted to 'man' by God. The remainder was strange, unknown and forbidden.[57] If European explorers had simply reached Asia by the western route, as Columbus seems to have believed, no threat would have been posed to such a world view. If a whole new land had been found, it would undermine ruling belief. Two documents appear to have played a vital role in changing the terms of the debate.[58] The *Cosmographiae Introductio*, published in 1507 by the Academy of St Die in France, argued simply that a fourth part of the orb had appeared, an addition that could be assimilated without difficulty. Published with the *Cosmographiae* was Martin Waldseemüller's 1507 map, the one to which Vespucci owes his fame. Waldseemüller hedged his bets on the crucial question of whether or not there might exist a sea passage to the east. The main chart showed a passage while an insert-map along the upper border allowed for no break in the American isthmus. But he did unambiguously show the Americas as distinct land masses, a vivid illustration of the new view of the world. The implications, although they took some years to become accepted, were revolutionary. Once the existence of a separate continental mass was acknowledged, the *Orbis* could no longer be restricted to its earlier confines. The entire surface of the globe was opened up for exploration. If more lands were found (including, potentially, the heavens) they too could be encompassed within the bounds of a more flexible universe without the need for prolonged ideological debate.[59] From the very beginning the exploration of the one frontier carried within it the conceptual seeds of others. Similar motives might also be involved in each case. Terrestrial colonialism is usually driven by economic imperialism. Far-seeing proponents of long-term space projects tend to have similarly exploitative intentions, however much disguised in a language of disinterested scientific advance.

The frontier is offered as a site to which pilgrimage can be made, whether at the remains of the Alamo at San Antonio (now upstaged by the simulacrum constructed for the John Wayne movie) or at hundreds of smaller shrines marked on the map along the highways and byways of the land. When the original western frontier was lost at home it was also shifted abroad, to a far-west that ran into the east, leaving further trails of destruction. From the western shores of California, the outer edge of empire moved across the map to China via islands such as Samoa, Hawaii and the Philippines, eventually to Vietnam and the rest of Indochina. This was less a novel development than a return to the expansionist and supremicist ideology used earlier to justify westward movement in terms of the 'manifest destiny' of the white race to spread around the globe from America to the Far East, carrying the latest phase of 'civilization' full-circle to the land of its supposed roots. The timing of the shift from domestic to overseas frontier discourse was not accidental. The social unrest that resulted from a series of sharp depressions in the last quarter of the nineteenth century was blamed by many on the loss of a frontier safety valve. More to the point, it was argued that the continued profitability of American business depended on the conquest of new frontiers, although this usually involved winning commercial rather than territorial ground.[60]

The drive across the Pacific can thus be seen as the second stage of a process that began with the voyages of 'discovery' to the Americas. In between the two trans-oceanic movements came an interval during which America was crossed, colonized, mapped, ordered and appropriated. The infinite extension of the frontier had in a sense been prefigured in early maps of America such as that by the Spaniard Diego Ribero in 1529. It depicted a western side of America that extended vaguely into an infinity that covered about a third of the map's surface. Later maps influenced the route and objects taken by the expanding American empire. The principal Pacific naval base and staging post was located at Pearl Harbor in Hawaii. This appeared to be the ideal location on the Mercator map used by Alfred Thayer Mahan, the great advocate of American maritime expansion. Hawaii appears to be ideally positioned midway between China and the west coast of the United States. An examination of the globe, or a polar projection, might have suggested otherwise and led to a different course of empire. A shorter alternative route might have been to the north-west, via the Aleutian islands off the coast of Alaska, a direction that would appear nonsensical on the Mercator and many other flat maps.[61]

The war in Vietnam was justified in terms of an arbitrary frontier line, the 17th parallel, which the United States claimed to be defending as a border between north and south, although it was meant to have marked no more than a temporary demilitarized zone. The Geneva agreement of 1954 said the line should not be interpreted as a political or territorial boundary. It was to have lasted only until 1956 when a general election would be held to choose a government for the whole of Vietnam (which would almost certainly have been won by Ho Chi Minh), rather than the legal fiction of the regimes supported by America south of the parallel. This is another example of the arbitrary line on the map that comes to have a decisive and in this case devastating impact on reality. Frontier imagery was widely applied by American forces on the ground: enemy ground was 'Indian country' and some military operations took the names of figures from the Wild West.[62] Vietnamese troops and civilians were routinely accorded the same 'savage' and inhuman status as Native Americans and were often subjected to a similar onslaught. Croplands were defoliated and rendered barren, robbing many of the ability to feed themselves in a tactic reminiscent of the destruction of the buffalo in earlier years.

In Vietnam many elements of frontier mythology were reworked, thrown into serious doubt, and in some cases revived for future use. With a familiar blend of evangelical mission and economic ambition, many Americans saw Asia as a new frontier landscape in need of enlightenment. The endeavour promised an engagement with reality, an escape from postwar decadence and complacency. This was one of the ingredients of Kennedy's New Frontier, a challenge to Americans to set aside selfishness and greed and help to invigorate a culture otherwise said to be in danger of lapsing into dullness and unauthenticity. Vietnam seemed to offer an ideal arena. America, it was claimed, could regenerate its own virtues while taking the fruits of civilization to the region. The United States could renew its mission, the original Puritan 'errand into the wilderness'. The tone of some commentators echoed the jeremiads issued by second-generation Puritans in the seventeenth century who had railed against the 'backsliding' of those deemed to have betrayed the initial promise.[63] Vietnam was to be turned into the kind of pastoral 'middle landscape' that had proved so elusive at home.[64] The combination of frontier mythology in general and the specific component concerned with a regeneration of authenticity offered a resonant ideological framework, coupled with Cold War axioms, within which to legitimize an imperialist endeavour. Philip Caputo recalls in his Vietnam memoir the pride and self-assurance of young Americans who decided to join the armed forces during the Kennedy era: 'We went overseas full of illusions, for which the intoxicating atmosphere of those years was as much to blame as our youth.'[65] Part of his reason for joining the Marines in 1960 was that he was swept up in this patriotic tide. But it was also a search for a more heightened reality: 'I was sick of the safe, suburban existence I had known most of my life.'[66]

Raised amid postwar affluence in Westchester, Illinois, Caputo recalls that the only thing he liked about his surroundings was an unspoilt forest preserve in which he hunted as a boy:

> There was small game in the woods, sometimes a deer or two, but most of all a hint of the wild past, when moccasined feet trod the forest paths and fur trappers cruised the rivers in bark canoes. Once in a while, I found flint arrowheads in the muddy creek bank. Looking at them, I would dream

of that savage, heroic time and wish I had lived then, before America became a land of salesmen and shopping centres.[67]

His wish appeared to be granted when he was sent to Vietnam, but the experience was a rude awakening rather than providing the sense of authenticity it had promised. A similar movement is found in Norman Mailer's *Why are we in Vietnam?* (1967), where the day before embarking for Vietnam a new recruit recalls his experiences on a hunting trip in Alaska's Brooks Range, 'the last untapped wilderness'.[68] There is an awareness here of the unauthentic nature of the war, evoked in hunting trips conducted with the aid of helicopter transport into the wilds. Yet the response of the two heroes is to venture off on their own without the aid of map and compass in a purification ceremony similar to that sought by the young Caputo.

The same kind of ritual engagement with a landscape of wild authenticity occurs in James Dickey's *Deliverance* (1970), a tale of four inhabitants of suburbia who head into the woods with their bows and arrows and canoes on a white-water expedition that turns into an elemental contest for survival. The narrator is vice-president of an advertising agency, his existence the epitome of dull image-bound routine. It is a last chance for the protagonists to pit themselves against the reality of the river. Its fast-flowing course is about to be dammed, reduced to the kind of bland undifferentiated expanse against which meaning is defended in Swift's *Waterland.* In this case it is not the natural state that is entropic but the social. The dam is a cultural imposition on the land, an appropriate metaphor for the expression of a mythology in which culture stands for a dulling of the edge of natural authenticity. The dynamic applied to the territory is prefigured on the map. A confusion between map and territory is found in the opening words of the book, the map itself becoming a hostile presence: 'It unrolled slowly, forced to show its colours, curling and snapping back whenever one of us turned loose. The whole land was very tense until we put our four steins on its corners and laid the river out to run for us through the mountains 150 miles north.'[69] An illusion of control is gained as Lewis, would-be survivalist and instigator of the project, marks out the course of the river with his pencil and gestures as if to erase the area to be submerged beneath the dam. Preconceived images continue to haunt the characters as the story develops. The narrator catches a glimpse of a reflected image that casts him in the role of forest man, explorer, guerrilla or hunter: images that seem to sustain him through the hardship to follow, just as they might have done for some of the American troops in Vietnam.

A central symbolic figure for Kennedy was the 'Green Beret', the Special Forces agent who was presented as part soldier, part missionary and development worker. He was to offer America a real and flexible counter-insurgency response to localized guerrilla wars of liberation, to replace the overbearingly clumsy and artificial threat of all-out nuclear conflagration. The Green Beret would be heir to the tradition of earlier generations of 'Indian-fighters' who had forsaken the niceties of formal combat structures and traditions in favour of an engagement with the enemy on its own terms, although genuine counter-insurgency strategies were opposed by the military hierarchy and rarely given much chance to succeed in Vietnam. Kennedy sought to identify himself with the Green Berets, championing the importance of their limited role against that of the regular army in the same terms as those in which footloose pioneers might be set against bureaucratic institution.[70] Such an identification was to prove fraught with difficulty, however, and to mirror the wider fate of the frontier mythology as it came into contact with the experience of Vietnam. If the Green Beret was celebrated in the American media as a new pioneer hero, he also had his dark side. He could appear to represent the worst of both worlds, destroying the natural inhabitant of the woods while at the same time corrupting his own values in an orgy of violence and destruction.

Concerted efforts were made to map and remap the territory, to bring it under control. Under French rule Vietnam had effectively been abolished on the map. It was divided into two protectorates, Tonkin in the north and Annam in the centre, with the south ruled as a direct colony under the name of Cochinchina. Along with Laos and Cambodia, the whole area was renamed Indochina and put under the control of a single authority.[71] Such divisions and regroupings were classic imperialist strategies, imposed in an act of reorientation and control justified by the claim that the three administrative areas reflected the existence of three separate peoples. Vietnam itself, we should note, was a relatively recent and artificial construct, invented in that name only in the nineteenth century.[72] Historically, the modern line drawn at the 17th parallel was only relatively arbitrary. It corresponded roughly with the boundary that for hundreds of years had divided what was to become Vietnam from the separate kingdom of Champa. The boundary was erased in the fifteenth century when the Vietnamese spread south. The 18th parallel came later to mark a division between two warring Vietnamese states as political cohesion was lost.[73] The real territory asserted against the impositions of modern imperialism was a more or less arbitrary construct, a fiction none the less real or worthy of defending for its lack of any ultimate grounding.

The American military command imposed its own map. The country was broken down into zones that took little or no account of the existing reality on the ground. Americans became 'cartomaniacs' in Vietnam, as Donald Ringnalda puts it: 'The maps provided the military with a reassuring ersatz reality, a reality traced from the imagination of the *American* landscape.'[74] The initial division of the territory into two might have appeared justified from a glance at the map. Viewed in the abstract the hourglass shape of Vietnam seems ripe for division along the narrow strip that links north and south. Closer examination of a relief map suggests otherwise. Vietnam owes its tortuous shape partly to the mountains that isolate the coastal ribbon. Some boundaries are relatively less arbitrary than others, although they remain far from necessary. American military maps depicted a network of posts across the territory that appeared to signify control, although much of the surrounding country remained in the hands of people sympathetic to the revolution.[75] In one attempt to 'pacify' the area around Saigon operations were conducted according to a series of concentric circles drawn onto the map across a zone thought to be large enough to prevent the enemy ever reaching the capital at its centre. On the map it looked orderly and logical enough, but in its equal and abstract expansion on all sides the schema failed to take into account areas of relative US or enemy strength and so proved largely ineffective.

Many fictional accounts of the war also sought to impose a rational, ordered mapping onto a seemingly amorphous territory. One of the most notable examples is John Del Vecchio's *The Thirteenth Valley* (1982), a work bursting with actual maps of its own, textual descriptions of maps and an often tedious density of statistical detail. Each chapter of the main narrative, describing a military incursion into the valley of the title, ends with a map showing the latest movements and positions. The map is accompanied by a report of 'Significant Activities', presented in block capitals and the cold, impersonal style of an official, supposedly objective, reckoning. The end of the book comes with a 'Final Tabulation' of the casualties as well as a glossary and a chart of key historical dates ranging from the founding of the First Vietnamese Kingdom in 2879 BC to the establishment of anticommunist opposition after the fall of Saigon in 1975. The histories of the characters are presented along with a variety of dialects in which they speak. One of the principals, Lieutenant Brooks, attempts to impose his own theoretical map on not just this but all conflicts – military, racial and personal. A range of discourses is thus brought into play in an attempt to bring events under control, or at least to make them representable and comprehensible. Brooks concludes that conflict is structured in cultural frameworks, particularly language. Yet there is little recognition that Del Vecchio's text is itself a manifestation of precisely the kind of relentlessly imposed cartography involved in the American war in Vietnam.

Not that any of these mappings did much to bring the territory under more than an illusory control. Jungle warfare against a rarely seen enemy was confusing, shapeless and almost impossible to map, as Caputo discovered: 'There was no pattern to these patrols and operations. Without a front, flanks, or rear, we fought a formless war against a formless enemy who evaporated like the morning jungle mists, only to materialize in some unexpected place.'[76] Hundreds of troops died in battles for hills marked only by a number on the map and that were often abandoned soon after being won. The Vietnamese wilderness was not to be forced into submission in a few years in the way vast areas of the American landscape seemed to have been tamed over the centuries. For foot soldiers such as Caputo the very ground under their feet, traditional home and security of the infantryman, became uncertain, the domain of mines and booby traps that would turn it into a deadly enemy. Like the Puritans before them, the troops were liable to react with disproportionate violence once their bearings were lost in the jungle wilderness.

When it was not liable to explode, the ground was often insubstantial swamp or engulfing quicksand. In Walter Hill's Vietnam analogue, *Southern Comfort* (1981), a patrol of Louisiana national guardsmen become lost even before their map is mislaid when their leader dies. The map is useless, unable to plot the shifting ground and the transformation into a watery channel of their intended route through a swamp, just as the frontier mapping failed to accord with the realities of the situation in Vietnam. In terms of frontier mythology, Hill's film effects a double projection: an attitude toward an enemy and its domain transposed from 'Indian country' to Vietnam is here replayed closer to the original setting as the patrol comes into conflict with a local Cajun community. Try as they might the protagonists cannot find their way to their intended target, the secure cultural imposition of order upon the territory represented by an interstate highway.

Where geometric order is imposed it proves either illusory or can offer only brief respite. One moment of passing order is captured in the battlefield game of checkers played by two characters in Tim O'Brien's haunting collection, *The Things They Carried* (1990):

> There was something restful about it, something orderly and reassuring. There were red checkers and black checkers. The playing field was laid out in a strict grid, no tunnels or mountains or jungles. You knew where you stood. You knew the score. The pieces were out on the board, the enemy was visible, you could watch the tactics unfolding into larger strategies. There was a winner and a loser. There were rules.[77]

Such rules and grids could not be maintained for long. For the Americans the Vietnamese landscape remained shifting and unstable, the anarchic jungle wilderness a constant threat, even within the military enclaves carved out by the occupying power. As Stephen Wright describes it in *Meditations in Green* (1983): 'There is growth everywhere. Plants have taken the compound. Elephant grass is in the motor pool. Plantain in the mess hall. Lotus in the latrine. Shapes are losing outline, character. Wooden frames turning spongy. The attrition of squares and rectangles. The loss of geometry.'[78]

For Fredric Jameson the kind of disorientation and bewilderment that resulted made Vietnam the first postmodern war, one that could not be expressed through traditional paradigms of the war novel or film. What Jameson has in mind is the form of representation resorted to in Michael Herr's *Dispatches* (1978), 'in the eclectic way its language impersonally fuses a whole range of contemporary collective idiolects, most notably rock language and Black language [. . .].'[79] Caputo might agree. Because of the sporadic and confused nature of the fighting, he says, 'it is impossible to give an orderly account of what we did.'[80] We are left with a series of disjointed and impressionistic fragments, a feature of many accounts of the war. Broader narrative frameworks were also reasserted, however, in attempts to make sense of the experience. Secularized versions of Puritan doctrine were available for those who clung to notions of disinterested mission. Other elements of frontier mythology were also replayed in almost all of the Vietnam narratives, whether ultimately they were undermined or reinforced. This applies far more widely than just to obviously crass examples such as John Wayne's rehearsal of the Alamo in *The Green Berets* based on the book by Robin Moore. Released in 1968, *The Green Berets* was one of very few accounts in which Vietnam was directly represented on screen before the end of the war, boosting the myth of its eponymous heroes at a time when their role in Vietnam had become insignificant.

A deliberate and highly successful attempt to resuscitate a mythology rooted in the frontier tradition was made just two years after the end of the war in the George Lucas film *Star Wars*. As John Hellman suggests: 'The reception of the original *Star Wars* film in the summer of 1977 certainly suggests that Lucas filled a painful gap left in American consciousness by the loss of the western and its frontier mythology during the Vietnam era.'[81] Not only an immense hit at the box-office, the film took on the dimensions of a major cultural event, spawning two sequels, a whole new era of science fiction entertainment and a huge industry of spin-off merchandise. It evoked both space and pioneer myths in a way that made its title the ideal label to be pinned onto Ronald Reagan's Strategic Defense Initiative in the following decade. In a highly suggestive reading, Hellman finds in the three *Star Wars* films a re-enactment of the fate of the frontier mythology as applied to Vietnam. In the opening chapter, he suggests, we have a straightforward working of the myth as understood before the war. *The Empire Strikes Back* (1980) represents the dark side of the myth, as the central figure Luke Skywalker impetuously takes on a force he is not ready to defeat and in the process discovers the dark side of his own nature. At the end of the second film, according to this reading, the narrative is at the same position as that reached in Vietnam epics such as *The Deer Hunter* (1978) and *Apocalypse Now* (1979): a moment of negative insight. But the *Star Wars* trilogy is able to move forward, eventually to a point of triumph. In *Return of the Jedi* (1983), we find Luke fully trained and ready to take on the dark side of 'the Force', 'armed this time with a power of self-knowledge that disciplines his youthful idealism, vigour, and natural skill into a mature – which is to say chastened – assuredness, command, and acquired craft'.[82]

It is hard to avoid the temptation to read this analysis into the situation as it applied to the chastened but assured architects of American foreign policy confronted by the situation in the Arabian Gulf in 1990–91. Western lore was certainly still available to provide orientation for those Americans reluctant to accept the need for the launch of a ground war to follow the weeks of aerial assault. President Bush gave Iraq a noon deadline to begin to leave Kuwait. That the deadline was noon *American* (Eastern Standard) time should not have surprised us. The mythic resonances were aimed at the American rather than the Iraqi people. The President did not have to utter another word for his ultimatum instantly to be translated by the media into the language of a *High Noon* in which he would unquestionably be playing Gary Cooper. The complex political situation could thus be reduced to the terms of a simple parable of reluctant good forced against its will to fight bad. A whole series of otherwise difficult arguments could be mapped onto a model flattering to the public conscience at a time when the American administration seemed to be doing its best to undermine Soviet peace initiatives that might have satisfied the principle conditions of the relevant UN resolutions.[83] The American response to the invasion of Kuwait was also articulated around an imaginary line on the map. 'A line has been drawn in the sand,' declared Bush in August 1990, inscribing onto the Saudi desert a demarcation that implied an otherwise fluid and expansionist Iraqi intention. The justification for war in both Vietnam and the Gulf also involved familiar pioneer rhetoric about the global mission of America to enter combat with supposedly dark and barbaric forces.

In the years between the wars in Vietnam and the Gulf, foreign policymakers had not been idle, applying Cold War doctrines similar to those articulated against Vietnam to Central and South American states such as Chile and Nicaragua whenever they appeared to threaten US regional hegemony. That the same ideological map was being imposed was demonstrated on one strategic chart of Guatemala. The country was divided into military areas the names of which included 'Saigon' and 'Hanoi'.[84] Ironically, it had been in Guatemala that a key piece of the future policy map had earlier been shaped. The socialist regime of Jacobo Arbenz was defeated with such ease by the CIA in 1954 (following its success the previous year in toppling Mohammed Mossadegh in Iran) that some were tempted to believe intervention could prove equally painless elsewhere. After Vietnam the United States returned to a policy of limiting itself mostly to fighting its wars by proxy, supporting right-wing regimes or opposition groups and not being led to the full-scale direct involvement of its own forces, a doctrine given fresh articulation by Nixon in the withdrawal of American troops from the front line in Vietnam.

As far as Vietnam was concerned, the *Star Wars* trilogy offered a highly revisionist account. With its tale of a small band of committed rebels taking on a vast Empire, the resonances were hard to mistake. Lucas himself identified the evil Emperor with Nixon, a casting which equates Darth Vader with Henry Kissinger or Robert McNamara and the rebels with the National Liberation Front. As Peter Biskind suggests, the wizened figure of Yoda thus becomes Mao or Ho Chi Minh and the forest-dwelling Ewoks in *Return of the Jedi*, defeating imperial forces with the use of sharpened sticks, bows and other 'primitive' technology, a more explicit analogue for the Vietnamese guerillas – 'Marin County-style'.[85] The wall of a soldier's room visited by Michael Herr was covered by a collage which included 'a map of the western United States with the shape of Vietnam reversed and fitted over California'.[86] What Lucas presents is, if not a reversal, a very one-sided mapping onto Vietnam of a range of west coast counter-cultural attitudes. This tends to obscure the more ambivalent possibility that this outlook – with its own search for an authenticity not present in modern industrial society – might have shared at least some aspects of the frontier imaginary that played a part in supporting the intervention in Vietnam. If the effect of the *Star Wars* films was to give a new lease of life to frontier mythology by suggesting that the good could overcome the dark side, others continued quite happily to plumb the more violent depths in the name of authenticity. Features such as *Uncommon Valor* (1983), *Missing in Action* (1984) and *Rambo* (1985), echoing the earlier Native American captivity narratives, argued against bureaucratic corruption and delay and in favour of direct physical action to rescue servicemen still allegedly held captive in Vietnam. It may be no coincidence that two of the film genres that resonated most loudly in American culture in the years after Vietnam were the childlike return to innocence suggested by the Ewok scenes in *Return of the Jedi* and developed with huge box-office success by Steven Spielberg in films such as *Close Encounters of the Third Kind* (1977) and *E.T.* (1982), and darker films such as *Rambo.* The two sides of the frontier mythology, the naively idealistic and the cruelly destructive, both seem alive and well.

Films of the post-Vietnam era have also reversed some of the key images that did much to galvanize opposition to the war. In *The Deer Hunter* we are repeatedly shown images of American prisoners of war with a revolver held at the temple during enforced games of Russian roulette, a strong but inverted echo of the infamous sequence in which a Saigon police officer was filmed summarily executing an NLF prisoner after the 1968 Tet offensive.[87] In the first Vietnam scene of the film a soldier is seen killing women and children in a sequence reminiscent of pictures of the My Lai massacre, although in this case the killer is from the other side. In *Rambo* the helicopters – ubiquitous images of the American presence in Vietnam – are flown by the Russians, while their quarry on the ground is a resourceful American rather than Vietnamese enemy. Bruce Franklin argues that this re-imaging was part of a conscious effort to rewrite the history of the war, to restore the discredited frontier mythology of heroic struggle against brutal enemies in order to build support for a remilitarization of American foreign policy in the 1970s and early 1980s. Deliberate or not, the inversion of such images could help to rebuild an aggressive pioneer consciousness.

Reopenings of the frontier abroad such as that which occurred in Vietnam had been foreseen by Frederick Jackson Turner, even as he declared the closure of the old frontier. It would be a rash prophet 'who should assert that the expansive character of American life has now entirely ceased.'[88] The west, as Turner was to write elsewhere, was 'a form of society, rather than an area.'[89] Thus, in Richard Drinnon's words: 'on our round earth, winning the West amounted to no less than winning the world. It could be finally and decisively "won" only by rationalizing (Americanizing, Westernizing, modernizing) the world, and that meant conquering the land beyond, banishing mystery, and negating or extirpating other peoples [. . .].'[90]

The eastern frontier opened up in Indochina was inseparable from the frontier marked by the 'Iron Curtain' and the perceived threat represented by the Soviet Union and China. These frontiers lost some of their ideological potency after the collapse of the communist bloc. Many in the West have welcomed these developments in a spirit of triumphalist vindication of their own hawkish stance. Others have sounded a more cautious note. Consciously or not, this may to some extent result from fears about the ideological consequences of the loss of a clearly defined and supposedly alien 'other' across the frontier. An assortment of alternatives or replacements have been conjured up over the years, including figures such as Colonel Gaddafi, the Ayatollah Khomeini, Saddam Hussein and Manuel Noriega. That some were once considered to be valuable allies seems to cause little difficulty to the process. The zone of threat has shifted rather than simply being erased from the map. Dark, hazy and unpredictable outlines are sketched in place of the simple binary opposition of the Cold War.

Within months of the fall of the Berlin Wall President Bush announced plans to put a man on Mars in the next 30 years, to open up a whole new frontier in space. *Life* magazine produced a cover story billing the red planet as 'Our Next Home', outlining vast plans for 'terraforming' Mars, using technology to transform it into a new arcadia. The first occupants would be due to arrive 'almost exactly four centuries after the Pilgrims landed near Plymouth Rock'.[91] In future centuries, the writers suggested, the new Martians might go on to colonize other planets, becoming 'the Americans of the interplanetary era'.[92] At first the settlers of Mars would live in self-contained 'biospheres' protected from the hostile existing environment. Their forerunners have already occupied a prototype biosphere in the Arizona desert, a space-age equivalent of the attempts of some settlers to construct their own hermetically sealed encampments within the American wilderness, a world which includes a rainforest tamed to the dimensions of a giant bottle.[93] In space the frontier can in theory be pushed back further and further, to the infinite regressions on the outer edge of the map of the universe. There is always some unknown against which a more familiar reality can be mapped, or to which appeal can be made for some new kind of authenticity. The line on the map around which articulations of the real and unreal can be articulated is likely to be an enduring one.

7 The Imperialist Map: Beyond Materialism and Idealism

[. . .] *I had a passion for maps. I would look for hours at South America, or Africa, or Australia, and lose myself in all the glories of exploration. At that time there were many blank spaces on the earth, and when I saw one that looked particularly inviting on a map (but they all look that) I would put my finger on it and say, When I grow up I will go there.*

Joseph Conrad[1]

The country was made without lines of demarcation, and it is no man's business to divide it.

Chief Joseph, Nez Percé[2]

If the view from space failed to offer any real escape from earth-bound perspectives as far as superpower relations and ideology were concerned, the same might be said of its impact on a broader philosophical debate. A writer in the Soviet journal *International Affairs* greeted the launch of *Sputnik* as a reaffirmation of the correctness of Marxist materialism, refuting 'idealistic theories that the world is "unknowable", that the objective world does not exist but is only a product of the human consciousness.'[3] Thus continued a debate as long as the history of Western philosophy itself. Materialist philosophy generally asserts that ultimate reality exists at the level of physical objects external to any human thought or understanding. Idealism counters that the pure actions of such objects upon the body can never be experienced because of the various ways either the sense organs or the understanding always shape or filter that which is perceived. Cultural cartographies cannot be reduced to either of these perspectives. In the socio-cultural sphere that we inhabit, reality is grounded only in the map-like grids of its own constructed landscape. On such a terrain it is impossible to make absolute distinctions between levels such as the base and superstructure of Marxist theory, however useful they might be as strategic tools in particular situations.

Various efforts have of course been made within Marxism itself to overcome the excessively one-sided materialism found at times in Marx's own work, some more successful than others, some seeking to point out that there is much more to Marx himself than a narrow economism.[4] Political factors, for example, may play an active part in the constitution of economic classes rather than merely reflecting class relationships given automatically in the workplace. It is well known that Marx distinguished between classes 'in themselves', defined objectively according to the positions of individuals and groups within the social relations of production, and classes 'for themselves' that had arisen to a state of class self-consciousness. A movement from the former to the latter seems essential to any revolutionary uprising. Political and ideological-cultural articulations are necessary if this is to be achieved. Political and other communicative processes are not a simple transmission of the pre-existing views of individuals or collectivities. They play an important role in the determination of such views. More directly economic factors such as the involvement of workers in strike action have a vital function in creating class self-consciousness, but political and ideological mechanisms are never absent. Political and ideological phenomena are also embodied in institutions and practices of a concrete nature rather than being restricted to the level of ideas. Economic and other forms are closely interwoven and all structurally implicated in a particular social formation such as capitalism, patriarchy, imperialism or state socialism.

In the settlement of Israel, the carving out of a Zionist state in Palestine, we see an example of ideology being converted into physical reality. Imperialism is a matter of territorial expansion and legitimation, as Edward Said says.

> A serious understatement of imperialism, however, would be to consider territory in too literal a way. Gaining and holding an imperium means gaining and holding a domain, which includes a variety of operations, among them constituting an area, accumulating its inhabitants, having power over its ideas, people, and of course, its land, converting people, land, and ideas to the purposes and for the use of a hegemonic imperial design; all this is a result of being able to treat reality appropriatively. Thus the distinction between an idea that one feels to be one's own and a piece of land that one

> claims by right to be one's own (despite the presence on the land of its working native inhabitants) is really nonexistent, at least in the world of nineteenth century culture out of which imperialism developed. Laying claim to an idea and laying claim to a territory – given the extraordinary current idea that the non-European world was there to be claimed, occupied, and ruled by Europe – were considered to be different sides of the same, essentially constitutive activity, which had the force, the prestige, and the authority of science.[5]

The power to map or to narrate, or to keep other forms of mapping at bay, is a key element in the ability to claim a territory. In many of the classic works of nineteenth-century English fiction Said finds a whole world of geographical relationships implicitly mapped out in a schema according to which other countries are no more than adjuncts to the Western commercial metropolis.[6] It is the very fact that these mappings are taken for granted in the texts that underlines their force and the currency of imperialist ideology at the time. The Israeli territory, likewise, would not have been sustainable on the map without the imposition of a firm ideological grid. Zionism was represented as the triumph of reason and idealism over an Arab culture defined only as its negative. Palestinian land rights were defined out of existence, like those of the Native Americans or other colonial subjects, a process that continued in the territories occupied in 1967.[7] The biblical narrative central to the Zionist project in Palestine was also a guide for some of those who settled the New World. The parallels may help to explain the particular resonance the Israeli cause has for many Americans today. The biblical mapping as applied to America is yet another example of a fictional (theological or ideological) map that imposed form onto a territory in advance of the exploration of its own topographical or other features. For some colonists, particularly the Puritans in New England, one of the basic dynamics of settlement was that prefigured in a selective reading of the book of Joshua: the command to cut down the forests, drive out the uncivilized Canaanites/Native Americans and build a new Jerusalem across the Atlantic.[8]

This whole trajectory of colonization was also founded in an important sense on a separation of the spheres of the material and the ideal, territory and map. Materialism in Western culture came to prominence in a complex series of conjunctions inscribed onto its own teleological map as the movement from a medieval to a Renaissance or Enlightenment world. If some thinkers sought an escape from the restrictions of unduly rigid theological certainties they often responded with equally dogmatic assertions of their own. Thus Descartes, freed apparently into the fresh air of reason, sought a new ground on which to plant his feet. Throughout texts such as the *Discourse on Method* (1637) and the *Meditations* (1641) these are precisely the metaphors that he deploys. His whole plan, he says in the third discourse, 'had for its aim assurance and the rejection of shifting ground and sand in order to find rock or clay.'[9] Descartes hoped to arrive at a point of irreducible certainty. What he found was simply a projection of the values of his culture: an insistence on the primacy of the existence of the individual thinking mind, a perspective that failed to comprehend the operation of mappings that transcend the individual dimension.

A metaphysical distinction between self and material world empowered geometers, geographers and explorers to treat new territories appropriatively. Rather than feeling implicated in the material world as a part of it, subject to an overwhelming sense of its vastness and capacity to bewilder, Enlightenment philosophers and their predecessors were able to embrace it in an all-encompassing grid. Michel Foucault charts this shift in terms of a movement from an epistemology based on resemblances to one founded on notions of taxonomic order, although as he suggests we draw a more or less arbitrary line ourselves when attempting to identify the point at which such changes occur. A system of correspondences in which heavens and earth, macrocosm and microcosm, are mapped mutually onto one another is replaced by an analysis akin to that applied by Linnaeus to the plant and animal world. Language is no longer viewed as something inscribed into the world itself but as a tool that can be applied to the world from outside.[10] A distance is established within which domination can be achieved. If this involved at one level a distinction between signifier and signified, it was a distinction that was also blurred again as the grids imposed by the colonial powers were naturalized and objectified. The entire globe could be mastered theoretically in a rational and abstract framework, even if much of it remained physically unexplored. When faced with the greatest geographical unknown of all, staring across the huge expanse of the Atlantic into the potential abyss that some believed led to the edge of the world, voyagers might be comforted and driven onward by the rationality of the grid lines of longitude and latitude that could bring it all into the orbit of a single spatial and cognitive matrix. This was a gradual process, of course, and we should not overstate the extent to which early explorers and seamen were familiar with the contemporary scientific thinking of a small intellectual elite. Religious rather than scientific truths remained dominant at the time of the great voyages of 'discovery'. The charts of the fifteenth century and most of the sixteenth were primarily medieval in character, as were the principal motives and assumptions of the early voyages. Spain and Portugal, the countries first to build empires in the New World, were steeped in feudal attitudes and institutions far removed from the intellectual ferment of Enlightenment circles in Italy.[11]

Eventually a new kind of space was mapped, abstract and Euclidean. When colonist encountered indigenous inhabitant it was on a ground as much theoretical as physical and one that had been mapped well in advance. The cartography in question was a potent blend of the scientific-rational and the theological, elements of Ptolemy, Descartes and Newton allied with Old Testament scripture. Enlightenment thinkers tended to believe in the existence of a single, transcendent truth and reality to which their maps corresponded more accurately than any others. This belief appeared to be confirmed by the unprecedented control Western science and technology helped to bring over other parts of the world, particularly as the industrial revolution provided the physical means to translate aspiration into reality. Western imperialism began to change the world to fit its map, seeming by an act of self-fulfilling prophesy to prove the map's superior grasp of the underlying realities of the universe.[12] Local particularities or rival mappings were generally ignored or rejected.

There were considerable regional differences between the colonial engagements of the various European powers. Greater accommodations to Native American reality were allowed in some colonies than others, or at different moments in the process of settlement. Early relations between colonial and native American societies were often symbiotic, the initial settlers largely dependent on the indigenous peoples for their survival.[13] The grid imposed upon the American landscape took little or no notice of existing mappings, however, even if some Native American terms survived in the names of colonial states and settlements. An illustration of the contrast between colonial and indigenous mappings can be found in the Southwest, at the one perfect point in the grid. The Four Corners Monument marks a place of mathematic perfection on the map of the United States, the only point at which four states (Utah, Colorado, Arizona and New Mexico) meet at right-angles. A place of no other interest or particular significance, Four Corners has as a result become a tourist attraction in its own right: a perfect case of the map preceding the territory. Anthropological reconstructions of earlier cultures in the area give a very different picture. The state boundaries of today are rigid and geometric. The earlier boundaries between the ancient Anasazi, Hohokam and Mogollon cultures were amoeba-like, amorphous and overlapping.[14]

Native Americans imposed their own mappings on the territory. Navajo tradition applied to the Southwestern landscape a cosmological map comprised of two dish-like structures representing the earth and the heavens in the figures of a man and a woman. The Navajo world was bounded by four sacred mountains that provided points of orientation. These were mapped at the domestic scale onto the structure of the hogan, four upright posts marking the cardinal points in a dwelling oriented towards the rising sun. The relationship between the symbolic points of reference and actual topographical features appears to have been ambiguous. There was a flexibility in the cartography that did more than simply reflect a physical landscape, allowing as it did for the possibility of territorial expansion. The Navajo looked to the stars for many of the frameworks that ordered their lives. Important dates were marked in the calendar according to the appearance of particular stars in the heavens, mapped on painted 'star ceiling' panels.[15]

Pueblo societies in the Southwest applied another set of mappings to the territory. In the traditional cosmology of the Keresan the map of the earth was square and flat. Particular colours, animals and spirits were associated with the directions of each of the four corners, a pattern typical of Pueblo culture. While their more mobile Navajo neighbours sought guidance from the relatively fixed patterns of the stars, the Pueblos oriented themselves to the movements of the sun. Both the ancient Anasazi and their more recent descendants appear to have aligned themselves in this way. They established their position in the cosmos according to mappings worked into the natural and built environment. Ritual calendars were created by tracing the lateral movement of the rising and setting sun across the features of the horizon. Strategic openings were created in buildings, carefully aligned to allow beams of light to shine through at particular moments to mark events such as the solstice or the equinox.[16]

Mappings were also imposed by nomadic groups on the open expanses of the Great Plains to the north. Cheyenne belief was oriented according to six powerful deities mapped onto the landscape. One was located above the earth and another below. The others were at the four points of a compass measured by wind direction and the daily movement of the sun. The stem of the pipe used in smoking rituals was pointed in sequence to each of the six directions. Similar orientations guided the Lakota who also applied a four-part schema to classifications of gods, directions, parts of time, animals and phases of life.[17] The dominant pattern of Plains life was circular, even if the circle was often divided into quadrants. The universe of the Mescalero Apache was a circle bisected along four cardinal directions. Four Grandfathers held up the world and were symbolized by the four principal poles supporting the Holy Lodge in which Apache rituals were held. The ceremonial lodges of the Lakota were circular but mapped according to a symbolic geography supplied by the territories of the winds blowing from north, east, south and west. The Lakota holy man Black Elk mourned the imprisonment of his people in the lifeless squares of modern reservation housing, an argument that might equally well have been applied to their entrapment in the grid of the colonial map. For Black Elk the circular world of the Lakota was attuned to the shape of the cosmos and of the terrestrial environment. The circle itself remains an idealized imposition on the territory, however; a mapping reflected in the ancient Plains 'medicine wheels', large radiating structures of stones oriented to both sun and stars. To Europeans equipped with notions of alienable property rights the nomadic tribes of the Plains might have appeared to leave no trace of tenure upon lands that were imbued with an array of spiritual mappings.[18]

The suggestion that Native American cultures are or were simply a part of or 'at one with' nature is the outcome of an ideological mapping that constructs the other as a single, distinct and 'primitive' ethnic group. The historical effect was no accident, even if the label 'Indian' itself was applied mistakenly as a result of the cartographical errors of European explorers who maintained that they had reached the east. The concept of ethnicity implied by the category was formulated in various ideological mappings, culminating in nineteenth-century theories of social Darwinism. An evolutionary development was proposed, beginning with 'savages' or 'barbarians' such as the Native Americans and culminating, predictably, with Western civilization. Precisely such a schema was mapped onto the American landscape by J. Hector St John de Crèvecoeur, who (like Thomas Jefferson) favoured the alternative of a pastoral agrarian idyll supposedly to be found between the two poles of western frontier and eastern metropolis. According to such mappings, Native Americans could conveniently be lumped together as natural or inferior beings legitimately subjected to removal, forcible change or extermination. As so often with such articulations, the label has also taken on a reality utilized at times by those oppressed according to its terms, either in alliances of indigenous groups against colonial encroachment or more recent assertions of Native American identity and rights.

It was the act of wholesale alienation and appropriation that made the colonial mapping so different, and so effective a tool. The map served both to fix and to segment the territory; to control it, to make claims of sovereignty and to package it for sale. Native American mappings were more complex and dynamic, even if in some cases their quaternary divisions gave them a superficial resemblance to the grid celebrated at Four Corners. The Navajo world was a multifaceted reality, constantly in process and with unstable boundaries resistant to any more than provisional fixing. Everything was interrelated. Much the same could be said of the Lakota, the Cheyenne and many other Native American cultures. There was no sense of the objectification inherent in the colonial map. Hopi tradition had no equivalent of Western notions of homogeneous and abstract space as a dimension distinct from all others, and hence alienable and subject to control. A different metaphysics was found, in some respects closer to the world view of Einsteinian relativity.[19] Individuals in Native American societies were usually seen as integral parts of a much larger fabric. They were linked intimately with the rest of the universe rather than being able to stand above or outside. The land was not a separate entity that could be alienated or sold. Native American notions of land rights were based on its use, as a benefit to the whole community and as a part of a cosmic whole, although territorial rights did exist among rival cultural groups. This is not to suggest that Native American cultures were necessarily benign. The great civilizations of the Aztecs or the Incas were based on imperial systems of domination over other groups, just as some North American cultures fell victim to others. All of these cultures had their own complex histories and politics rather than occupying some timeless natural domain. The comparison to be made is between one particular kind of cultural cartography and others rather than between mapped and unmapped, culture and nature, or any such essences.

The Western colonial map is an abstraction that tends to extinguish other dimensions of reality in an act of violent appropriation. The degree of estrangement from the physical landscape may have been unique. The colonial grid has been implicated in a history of environmental despoliation perhaps never before seen on the same scale; or, more to the point, never conducted as an act of principle founded on the assumption that human improvement was based essentially on its opposition to the world of nature.[20] The contrast can be seen in the different attitudes of native and colonist towards precious stones and metals. The natives of the New World valued them for their beauty but saw them merely as parts of a larger design. The Europeans sought gold and silver for their worth as commodities abstracted from their context, as can be seen from the writings of Columbus onward. The Oglala Lakota knew of the existence of 'yellow metal' in the sacred Black Hills of South Dakota. It was of no particular significance to them, although gold-fever was to lead the colonists to breach the treaty agreements according to which the Lakota were supposed to have been guaranteed permanent settlement. The spiritual landscape of the Lakota has subsequently been mapped and remapped in an ongoing process of alienation. From 'Indian' reservation, it became a rich source of gold and later uranium. At the same time it was produced as a tourist site. Onto the map were inscribed the former battlefields of the Little Big Horn and Wounded Knee. Into the landscape at Mount Rushmore were carved the faces of Washington, Jefferson, Lincoln and Theodore Roosevelt in a supposed celebration of democracy. Frontier constructions and sub-Disneyland attractions abound, drawing visitors by the thousand.[21]

An increasingly mechanical and quantitative mapping was imposed by 'Enlightenment' thought and the impact of capitalist economic dynamics. The world was rendered orderly by being broken down into isolated compartments. Emphasis was put on the aspects of reality most suited to mathematical investigation. Other areas of experience were devalued.[22] They became less real, according to scientific doctrines, although supposedly rational analysis remained implicated in a range of other more complex and ambiguous mappings, including the religious. The same might be said of the colonial view of native peoples. The spiritual dimensions that accounted for many of their cartographies were to the invaders invisible, unmeasurable and therefore non-existent. In the face of the onslaught that resulted 'the little worlds of the primitives [*sic*] shattered as a Micronesian bamboo map might in a hurricane gust.'[23]

On the colonial map the indigenous presence was effectively negated and written out of existence. Rather than in any serious or lasting way being respected in their own right, Native American territories were seen generally as empty spaces on the map into which the settler could move and take up ownership. This was no accident. The promise of access to 'free' land was the single greatest motivation for many of those who crossed the Atlantic, whether as a source of profit for the wealthy or private subsistence for those escaping feudal tenure in Europe. That the peoples already occupying these lands had no concept of permanently alienable private ownership made the process seem all the more easy. Even if their presence was considered the indigenous peoples were not felt to be in proper possession in the first place. Western notions of land use were applied in an ideological mapping that declared Native American practices inefficient, wasteful and hence invalid. Pseudo-legalistic devices were also developed to deprive them of their land in ways more subtle and less expensive than the use of military force.[24] The remnants of Native American groups were remapped onto the landscape within the confines of officially sanctioned reservations. They were often subjected to further removals, erasures from the map and reinscriptions elsewhere as former reservation land was sought by white settlers. As the geometric outlines of the reservations suggest, the aim of official United States policy was to assimilate their occupants to the ways of the dominant power rather than to recognize their difference. Initiatives such as the Dawes Act of 1887 were used to break down communal tribal ownership systems and to divide the land into individual allotments.

The remaining reservations feature prominently on many of the most popular representations of the territory today as maps of 'Indian Country' produced for the tourist. A map of precisely that title is available from the American Automobile Association, covering parts of Utah, Colorado, Arizona and New Mexico. When they had been defeated and imprisoned in the colonial grid the remains of Native American cultures could be acknowledged and even celebrated for their authenticity. Again we see the flexibility with which notions of the real can be articulated around either side of the frontier boundary. On the eve of the 500th anniversary of the first voyage to the Americas by Columbus, *National Geographic* offered its readers a simulacrum of an 'America Before Columbus' that seemed more authentic than that inhabited by the modern American. It included impressionistic accounts and illustrations of life in a variety of native cultures. A narrative purporting to describe a Mohawk village in 1491, the year before the first landfall of Columbus, was presented in a present-tense description as if to offer to the reader an illusion of participation. Accompanying the issue, inevitably, was a map of 'Native American Heritage' across which were inscribed existing reservations, tribal centres, museums and archaeological ruins to which were guided tourists wishing to pay homage.[25] Once written off the map, Native Americans were allowed to return, but only as a shadow of their former presence.

Much the same process is found in changing conceptions of the American wilderness, the threatening landscape that often occupied the blank spaces on the map. It was only when wilderness seemed to be threatened with extinction that it began widely to be valued. In the early years of the republic wilderness remained for most Americans something to be conquered, an alien space that threatened the newly imported colonial reality. Appreciation of its aesthetic qualities often came less from those who had to interact with it on a daily basis than from the occupants of cities. From the comfort and safety of the city, members of the literary elite laid the foundations of a new outlook that found something to admire in the wilderness landscape. Concepts of the sublime and the picturesque were combined with a primitivistic idealization of what appeared to be a more authentic life closer to nature.[26] Transcendentalists like Emerson and Thoreau found in these landscapes the possibility of embracing the world afresh, with a clear eye untainted by the cultural inheritances of the past. From the latter part of the nineteenth century, when the first national park was established at Yellowstone in Wyoming, some of these notions were translated into a wilderness preservation movement that ensured the protection of wild enclaves in much the same way that reservations guaranteed some remaining space for Native American groups. Bounded and mapped as they were, these remnants of wilderness were shorn of their capacity to bewilder.

No such recuperation has been possible of the wilderness as experienced in the jungles of Vietnam, precisely because it defeated all of the best efforts of American cartography. A telling comparison between the two kinds of wilderness landscape is made in *The Deer Hunter.* The clean, crisp and bright mountainous terrain of the Appalachian mountains is accompanied in the film by the strains of devotional music. The mood is similar to that evoked in the opening pages of the James Fenimore Cooper novel, *The Deerslayer* (1841), from which the film appears at least in part to take its inspiration. A possibility of some kind of communion is suggested for the central character, Michael, who seeks to escape from the hypocrisies of his steel-town community. In the mountains Michael attempts to abide by a strict code and a proper way of hunting very like that of Cooper's wilderness hero Natty Bumppo. The Vietnamese jungle is presented in startling contrast as dirty, messy, confused and disorienting.

Where they have been made safe for consumption, wilderness landscapes are highlighted on regional maps of the states and closely mapped themselves for the benefit of tourist-pilgrims. Monument Valley, the place Baudrillard saw as offering a metaphor for a process of cultural desertification, has met no such fate as far as tourism is concerned. Visitors are supplied with a mapped route that enables them to drive in and out of the mesas and buttes. Onto these features have been mapped a variety of names and images that appear to have been imposed with little regard to native realities. Yet sites such as this seem to offer a point of contact with something more authentic than the business of modern life. In a nation often erroneously viewed as being devoid of ancient historical ruins[27] pilgrimage can be made to Monument Valley, where indeed the massive outcrops of rock rise out of the desert floor like the ruins of a great civilization. Close up, the smoothly sculptured facade of one butte could be the end wall of some vast cathedral, and is capable of provoking a similar sense of awe. Centuries of erosion have picked the surfaces of others out uncannily to resemble the remains of ancient columnated structures.

Along a stretch of Interstate 15 near the outpost of Barstow in the middle of California's Mojave Desert the tourist is offered a choice or combination of three attractions that express different elements of the interplay between reality and unreality, the authentic and the fake. First there is the real-fake: Calico Ghost Town. This is a old mining settlement restored with a blend of original remnants, reconstructed period-style shops and modern souvenir consumerism. A walking tour through the old mine workings is offered to those wanting to come a little closer to the underlying reality of nineteenth-century mining. Just a few miles to the north the visitor unsatisfied by the experience of the ghost town is led to the real-authentic: the Calico Early Man Site, an important archaeological site first discovered in 1968. Here the tourist is supplied with a self-guide map and tour of the excavations of a settlement thought to date back some 200,000 years, a chance to dig back to a truly primordial ground of human existence in the continent.

The third attraction to which visitors are drawn by map and guidebook is the colourful rock formation of Rainbow Basin, north of Calico, the outcome of some 30 million years of erosion. Like the Early Man Site it offers an expression of ancient authenticity, with its fossilized animal remains and the spectacular exposure to view of the clearly defined strata of eons past. This of course is only a side-show compared with the Grand Canyon, into which the visitor can descend through layers of rock in what seems like a journey into the primeval mists. A map for this trip is conveniently mounted on a display board at Pima Point, on the Canyon's south rim, charting the different strata eventually underpinned by metamorphic structures dating back some two billion years. Alternatively the canyon can be experienced in high-definition simulation on the screen of a giant IMAX cinema a few miles back from the edge. A 34-minute show promises the discovery of 'a Grand Canyon that would take a lifetime to experience'. Sites like the archaeological dig at Calico and geological formations such as Rainbow Basin or the Grand Canyon may seem to have little in common with the crude commercialism of the ghost town, but in each case a similar dynamic may be at work for the visitor: that curious displacement according to which real and fake, the experience of ancient authenticity and contemporary copy, become intermingled and mutually implicated.

The ultimate destination on Interstate 15 for the connoisseur of the dialectic between the real and the unreal (postmodern or otherwise) is of course Las Vegas, that glitzy outpost of neon facade set in the desert. Cliché or not, the city remains perhaps the finest expression of American unreal-reality, or as good a spatial metaphor as those found in California or Florida for the cultural map imposed on the void. Windowless, clockless and open for business 24 hours a day, the casinos of Las Vegas offer an entirely artificial environment, abstracted from any 'natural' rhythms of space and time. It is a landscape of dreams, yet the fantasies of wealth are framed by very present realities of financial loss. It is a place where ties are loosened, where marriage and divorce come easily; where productive economy gives way to gaming, but where gaming is the basis of its own real economy. The texture of the city is artificial, attractions in the casinos including a fake exploding volcano and ersatz Romanesque. Yet they also revert to assertions of authentic reality in order to compete for custom. Thus the high-wire act flying above the clatter of the slot-machines in Circus-Circus is the authentic article, a genuine Italian troupe doing it for real. On the road to Las Vegas the billboards include invitations to visit the dolphins at the Mirage casino. On arrival, confronted with the stone *statues* of dolphins mounted outside, the visitor could be forgiven for believing that the sign has outdone the reality in a cheap marketing trick appropriate to the name of the establishment. This is a common experience in the American landscape. Many of the franchised diners alongside the interstates are flimsy, cheap and tacky, vastly outweighed by the enormous signs announcing the distance to and attractions of the next outlet. As it happens this is not the case with the dolphins of the Mirage. Inside the casino there are indeed real dolphins on display, disturbingly, in addition to a pair of rare white tigers.

The geological landscapes of Rainbow Basin or the Grand Canyon might offer another metaphor for the cultural cartographies we inhabit. Cultural mappings may have no absolute ground, but then neither does the physical stuff upon which we live. Standing at the edge of the Grand Canyon we might be encouraged to doubt the ultimate solidity of the surface across which we work our mappings, although it still seems firm enough to support our weight. Even looking into the abyss it is difficult to absorb the fact that beneath all this, under the earth's crust, is a seething, aqueous magma. Neither is it easy on any cultural ground to appreciate the ultimate provisionality of its own structures. We can theorize the 'magma' of the cultural landscape, as it were. It is possible to conceptualize the underlying instability, the threat of destruction or dissolution, but it is doubtful that we can live it. We may experience moments of revolutionary upheaval, just as those who live above geological fault-lines are on occasion obliged to remember the precarious nature of their existence. If the ramparts of the elevated US highway were a suitable metaphor for cultural mapworks on the territory, the fate of the double-deck section of Interstate 880 approaching the Bay Bridge in San Francisco, flattened in the earthquake of October 1989, may be a telling image of their potential fragility. New mappings can always be constructed, however, along with new buildings and roads that soon take on their own appearance of solid reality.

The supposed access to the real offered by latter-day encounters with Native American culture or wilderness landscapes is itself a product of a particular set of conventions associated initially with the Romantic movement of the nineteenth century. After the achievement of political independence a search began for specifically American artistic material with which an equivalent high-cultural nationalism could be asserted. Romantic conventions imported from Europe provided a framework within which the Native American and the American wilderness could be elevated to artistic respectability.[28] Romantic celebrations of Native Americans or their environments were used by some American critics in the nineteenth century to attack the materialism of their own culture in terms of its artificiality or inauthenticity. Similar arguments were also deployed during the war in Vietnam. The image of the 'Indian' has been used in many different ways, but always primarily for what it signifies in the terms of others.

Romantic assumptions also seem to underpin Daniel Boorstin's emphasis on the distinction between the settlers and the European world from which they came. In opposition to grand theoretical constructs such as those of Newton and his contemporaries, Boorstin places the more humble everyday achievements of Americans who catalogued the flora and fauna of the continent. The argument is founded on a simplistic interpretation of the move to the New World as an escape from centuries of accumulated cultural baggage, an emergence into the open air from a cluttered landscape in which the bare earth 'was almost nowhere visible'.[29] Differences there might have been, but they did not amount to anything like an epistemological break. Boorstin's case rests on a circular appeal to 'self-evidence' as the basis of a new American concept of knowledge, canonized in the Declaration of Independence. Americans, it is suggested, could make their own maps, based purely on their own experience of the territory and unencumbered by any theoretical inheritance. The supposed objectivity of their findings is said to be validated further by the fact that they were expressed in the everyday language of the ordinary people rather than being couched in obscure scientific terminology. But everyday language is nothing if not a repository of sedimented cultural meanings. It cannot be appealed to as an arbiter capable of anything other than an instant confirmation of its own values. There is perhaps nothing we should trust less than appeals to self-evidence or common sense. They shift onto the ground of the natural, straightforward and non-dogmatic that which can only ever be the product of particular mappings. What Boorstin really describes is an objectifying (rather than objective) cultural cartography that in its very disavowals gains its greatest power to encompass, articulate and appropriate American territory, both physically and conceptually.

A more revealing insight into the epistemological assumptions underlying American policy comes from Henry Kissinger, National Security Advisor to Richard Nixon from 1969 to 1972 and Secretary of State in both the Nixon and Ford administrations. Kissinger is quite clear that there is an epistemological gulf separating the West and the 'Third World', a gap owing to the failure of the latter to undergo the equivalent of a Newtonian revolution. In Kissinger's view this is a weakness on the part of others rather than illustrating the possible shortcomings of a narrowly rationalizing Western epistemology. The West, Kissinger argues, 'is deeply committed to the notion that the real world is external to the observer, that knowledge consists of recording and classifying data – the more accurately the better.'[30] 'Third World' cultures 'which escaped the impact of Newtonian thinking have retained the essentially pre-Newtonian view that the real world is almost completely *internal* to the observer.' Implicit in this account is the suggestion that the external reality inhabited by the West is a superior and objective one. Others live out some kind of illusion. American intervention in the 'Third World' thus 'not only brings technology and consumer goods into play,' as James Gibson puts it, 'but also brings *reality* to the Third World. In claiming the West's radical monopoly on knowing reality, the Third World becomes *unreal*.'[31]

Yet the map that was imposed on parts of the world such as Vietnam during Kissinger's tenure and before was itself steeped in unreality. The American engagement in Vietnam involved no greater comprehension of the alien 'other' than the practices of Columbus and his successors in the New World. When Columbus made his first landfall in 1492 he immediately assumed the local population to be inherently inferior, largely on the basis of their lack of Western technology. He failed to appreciate sophisticated native methods of agriculture and government that were finely adjusted to suit the territory, just as the American politico-military establishment was unable to comprehend the particular factors that would enable the supposedly 'primitive' enemy in Vietnam to defeat the technological might of a global superpower. In both cases Western rationalizing projects myopically projected their own mappings onto other cultures. The contributions of technological progress and scientific management to American success in the Second World War helped to establish an increasingly narrow mechanistic view of the world. Kissinger himself claimed that little was beyond America's technocratic capabilities. American foreign policy since the war was based on the assumption that 'technology plus managerial skills gave us the ability to reshape the international system and to bring domestic transformations in "emerging countries".'[32] The scientific revolution had 'for all practical purposes, removed technical limits from the exercise of power in foreign policy.'[33] Many continued to believe that America could control the way in which the map of the world was redrawn as former colonial states gained their independence in the postwar era.

Such a belief played its part in the decision to intervene in Vietnam, to ink in and 'defend' the line across the 17th parallel that was supposed to keep communism at bay. American strategy owed more to preconceived maps than to any detailed consideration of the territory. Robert McNamara, Kennedy's secretary of defence, had developed techniques of systems analysis for use by the Army Air Corps in the war against the Axis powers. Under Kennedy he established a systems analysis office in the Department of Defense that worked on scientific analysis of potential warfare scenarios in the belief that all could be reduced to rational and quantitative terms. These were the latest, computer-aided developments in a long history of attempts to apply formal models to the conduct of war. Stylized map-boards were used by the Chinese to learn tactical manoeuvres from about 3000 BC. Similar games were developed in India. It was amid the other developments of the Enlightenment, however, that the modelling of warfare began to come into its own. Along with most other things the conduct of war was assumed to be subject to scientific laws that could be studied, detected and applied on the battlefield.[34] Models were widely used from the early eighteenth century, particularly in France and Germany. Developments in gaming often coincided with advances in the map technology that provided the principal theatre of operations. Intended primarily for abstract training in matters of strategy and tactics, battles fought out with counters on maps came increasingly to be used as a basis for the planning of actual campaigns. Schlieffen blamed deviations from his war-game based plan for the German invasion of France when it failed during the 1914–18 war. But the plan itself had not taken into account a number of key factors, particularly the likelihood of British intervention if German troops went through Belgium in an attempt to outflank the French.[35]

American war-gamers of the twentieth century sought to learn from such shortcomings by adding political factors to their reckoning. The Cold War Division of the Joint War Games Agency attached to the Joint Chiefs of Staff probably explored in advance many of the crises that racked the Kennedy and Johnson administrations in the 1960s. Two incidents that were not prefigured, however, were the Cuban missile crisis and the construction of the Berlin Wall.[36] Scientific models are unable to take into account all possible scenarios, no matter how many options and variations can rapidly be played out with the latest computer technology. Computer gaming may have given an impression of high-tech rationality, but it was essentially just another expression of a particular cultural perspective, all the more dangerous for its aura of objectivity. The more apparently realistic such models or virtual worlds become the greater is the danger that their mappings will be imposed onto the real world.[37] The models applied to Vietnam were inappropriate. Vietnam was seen as little more than a pawn in a game being played by the two superpowers. Local and historical realities were largely ignored. The confrontation was seen as a mechanical problem or a bargaining situation, 'rather than a struggle with its roots in ancient feelings of patriotism, desire for justice, and resentments of foreign intervention that might not respond to a "rational" challenge.'[38]

Shared by many academics and policymakers, Kissinger's Newtonian vision was founded on a mechanistic view of the world, a 'social physics' as Gibson terms it, that encouraged the idea that any problem could be solved by the application of sufficient force. When the French were defeated at Dien Bien Phu in 1954 the triumph of the Vietminh was based on social factors, relationships between soldiers and peasants that meant thousands could be mobilized for the assault. The land reform of 1953 played a key role.[39] Mythic figures of heroic resolve from the Vietnamese past were also used at various times to encourage resistance. The American administration failed adequately to understand the power of such factors. The answer provided by social physics was that more force was required, that the French failed simply because they lacked sufficient military strength or will. As Gibson puts it: 'What is at issue concerns conceptually mapping "nature" onto society, [. . .] rendering the social world invisible.'[40] Metaphors such as the 'domino theory' of Soviet advance or the 'chain reaction' among supposed satellite states came from precisely this abstract, mechanistic view of the universe.

Removed from the realms of history, politics or ideology, Vietnam was conceptualized as a laboratory. For Kennedy it was a place to test the requirements of guerrilla warfare. For the weapons manufacturers it was a testing ground for their latest hardware, while the armed forces were able to experiment with new aspects of military science. For certain politicians and social scientists Vietnam was a laboratory for 'social systems engineering', attempts to create economic modernization according to schemas such as Walt Rostow's ethnocentric 'stages of growth'.[41] Where possible the territory itself was converted into abstract spaces. Many of the inhabitants of rural areas were relocated from their villages into 'strategic hamlets', fortified constructs designed to remove people from the influence of the enemy. The main effect was to increase opposition to the South Vietnamese regime by threatening a culture based on ties to the land. Other areas were declared 'free-fire zones' in which anyone who would not abandon their home was declared a self-defined 'legitimate target' for massive bombardment.

These and other targets were often selected randomly from grid coordinates on the map. Bombers and artillery fire were directed into areas where knowledge of the territory was limited. Many raids were flown for little purpose other than the maintenance of mission quotas. Naval and air force chiefs competed against each other for increased quotas and a greater share of the military budget. Yet another arbitrary mapping was imposed on Vietnam as the territory was carved up into separate zones designated as the exclusive preserve of each service. Hanoi and the area immediately north of the demilitarized zone were allocated to the Seventh Air Force while Haiphong and north-eastern North Vietnam became the property of the

Seventh Fleet. 'The commander in Saigon could control only targets in the extreme south of South Vietnam, where the Seventh Air Force reported to him. The flights of the B-52s were separate from all this because they were controlled by the Strategic Air Command.'[42] The purpose of this mapping was not to improve operations but merely to ensure that each service could properly be credited for the number of missions undertaken.

Unable to grasp the reality of the situation on the ground, the American military sought to destroy it, to bomb and defoliate forests and undergrowth in an attempt to impose its own order on the land: 'They tried to create a physical terrain equivalent to the abstract mathematical space of 1,000 meter by 1,000 meter grid squares necessary for jets and artillery to find orientation.'[43] The effect was usually counter-productive, converting more of the people to the NLF. By the end of the war much of the rural population of the south had been remapped into towns and cities that had little productive base. The result was social and economic chaos rather than the order and control envisaged by American planners seeking to destroy the rural base of guerrilla wars of national liberation.[44]

One of the most extreme manifestations of the desire to rationalize the territory with the aid of the latest technology was the concept of the 'electronic battlefield', championed by General William Westmoreland. A network of sensory devices and listening posts was to have been used to create an electronic barrier south of the demilitarized zone to keep out the alien 'other'. The plan was abandoned on the grounds of cost, but a similar scheme was put into practice in the southern part of Laos, another victim of illegal American interventions. Thousands of electronic monitors and sensors, disguised as plants, were dropped into the jungles on the Ho Chi Minh trail, the complex network of routes used by the north Vietnamese to move troops and supplies into the south of the country.[45] Electronic ears, seismic devices and infrared sensors capable of measuring local changes in temperature were used to relay masses of data via aircraft circling overhead to an American computer complex in Thailand. When sensor readings reached threshold levels thought to indicate the presence of an enemy convoy a target would appear on an electronic map monitored by a computer operator. The whole of southern Laos was mapped into pre-selected sets of map coordinates divided into sections under the control of different technicians. Computer and radio links to American bombers enabled their navigation systems to be guided to the chosen point on the map to drop their loads either manually or on automatic command from the ground. The system seemed to promise the ultimate achievement of the mapping of scientific rationality onto the theatre of war, although in practice it proved rather less certain. No one could really be sure whether it had any significant impact on military realities at all. The sensors were destroyed in the bombing and so unable to record the aftermath. They could often be identified by the enemy or their local allies and decoyed by recordings of non-existent convoys to lure the American bombers away from the area of real operations. How many real convoys were hit, and how effectively, was not known.

Where social structures were recognized in Vietnam, rather than being assimilated to abstract mechanics, the Americans sought to change them. Various efforts were made to impose the Western-style mass-consumption society that was considered by Rostow to be a high stage of development. The real aim was to break down older loyalties and attachments where possible and to create new consumption communities akin to those established in the America of the late nineteenth century. American propaganda leaflets and broadcasts spoke only to American values, failing entirely to engage the specific realities of Vietnam. It is hardly surprising that many of those who served on the American side described the experience as unreal. The American bureaucracy tended to believe its own propaganda, confusing a 'self-promoting paper universe with the lived reality of [the] rural Vietnamese'.[46]

American troops on the ground were little better served. Senior officers often gave their orders from the perspective of an aircraft flying a safe distance above the ground. From such a viewpoint they could easily become detached from the tangled reality below. Some demanded movements or assaults that might have seemed simple enough from the map-like view spread below, but were dangerous or impossible from the middle of a jungle or a swamp. The postwar American faith in technological solutions was closely linked to this abstracted view from above. The technology of superior air power seemed to offer the prospect of almost painless victory, an escape from the high human cost of prolonged engagements on the ground. American advocates of air power in Vietnam believed or claimed that the war could be won by massive bombardment alone. They ignored the lessons of the 1939–45 war and the conflict in Korea, as well as the contrary analysis provided at the time by the CIA.[47] The myth of airborne omnipotence was revived by the overwhelming success of American air power in the Gulf war, despite the very particular nature of a desert conflict that might not be the best source of models to be applied elsewhere. The clean, non-entangling promise of air strikes was again offered at various instances as a solution to the problem of intervention in Bosnia-Herzegovina in the early 1990s.

In Vietnam the principal reality for many members of the officer class belonged to the written records and charts that appeared to map the gradual success of the American project. Progress was measured in body counts and promotions among junior officers were decided often by abstract kill-ratios and how many missions their patrols had completed, although the figures were often rejected as inflated by the Pentagon systems analysis experts. Lives were risked unnecessarily to mount some patrols that, even within American military logic, served no purpose other than the accumulation of statistics. Many soldiers died after being ordered immediately back into areas in which they had been fighting in order to find and count enemy bodies. Some refused to do so and invented the figures, which tended anyway to be exaggerated and unreliable. Illusions were constantly reported as fact and believed by those higher up the chain of command. Assessments of the area under enemy control were indicated in red on the transparent overlays placed over military intelligence maps. If the amount of red appearing on the overlay did not accord with the official version it would have to be reduced, regardless of the reality.[48] Those who insisted on the reality on the ground risked damaging their career prospects.[49] The complacency of many Americans was shattered by the Tet offensive of 1968, in which the National Liberation Front attacked dozens of towns and cities and briefly held part of the American embassy in Saigon. The offensive was a failure in immediate military and revolutionary terms. Many of the NLF's most experienced cadres were among the thousands killed, while the attack failed to spark an uprising of the urban population. The symbolic and psychological impact in

America was enormous, however, reflecting the twisted logic of a war in which appearances often seemed to outgrow realities.

Paradoxes such as these laid bare the irrationality at the heart of the 'rationalism' that was supposed to have guaranteed victory. The mapping that was brought to bear upon Vietnam failed in most aspects. It failed to give a clear understanding of the enterprise, while the omnipotence envisaged by Kissinger proved illusory. A rational, technological approach provided neither an accurate map of the territory nor the ability to defeat the enemy. The abstract spatial matrix imposed on the territory was neither entirely real nor purely imaginary in its impact on events. This was, as we have seen, nothing new in the history of America, its own settlement having entailed similar dynamics. In seeking to explain the American defeat many military figures, politicians and other commentators continued to repeat the same illusions. The Vietnam America had entered was characterized, variously, as a 'swamp', a 'quagmire', a 'morass', a 'mire' – indeed, Boorstin's 'thicket of unreality' – into which the United States had been led as the result of well-intentioned mistakes. A war founded on deeply rooted notions of 'rationalism' and into which America came for reasons of global policy was thus translated onto another plane: 'The war becomes conceptualized as some place *beyond* rational conceptualization [. . .].'[50] This is a danger also for those who over-emphasize the fragmentary, surreal or 'postmodern' quality of the war. The official mapping may have proved illusory, but the experience was far from unintelligible. To suggest that it was is to risk complicity with the apologist view that the war was an aberration rather than an embodiment of central aspects of American culture with parallels in both past and future. A similar element of complicity may be found in advocates of 1960s counter-cultural movements that sought precisely such a location 'beyond rational conceptualization.' For some the counter-culture offered an alterative to the alienated map/territory distinction of the Enlightenment. In place of an exploitative pioneer experience, some of its participants identified with the 'others' represented by the Native American or the Vietnamese peasant. Different forms of consciousness were embraced through the use of mind-expanding drugs or an attraction towards Eastern religions. The counter-culture also shared with its apparent opposite an abiding concern with the achievement of an 'authentic' rather than culturally mediated experience. The form and content of some narratives were challenged or undermined by the experience of Vietnam, but they could also be reconstituted in other guises, new cartographies, a process that often involved a blurring of conventional distinctions between fact and fiction.[51]

The fact and the fiction of the Vietnam war, the rationality and the irrationality, are not easily separated out. The domino theory contained an element of truth, however deeply enmeshed in Cold War fantasy. A successful socialist or communist regime in one post-colonial state might indeed become a model for others, however different the local realities. Vietnam was of little importance in itself to American imperialism, but it was located in a region of strategic concern. America wanted Southeast Asia to remain a part of the capitalist world economic system. The immediate issue was the future orientation of Japan, which became the linchpin of American policy in the Pacific following the 'loss' of China to communism in 1949. Southeast Asia was a vital alternative source of markets, raw materials and rice for Japan if it was not to be tempted into a regional alliance with China from which the United States might be excluded.[52] An abiding theme of American policy was the fear that the globe would break down into a series of separate trading blocs. The United States favoured a largely open world economy that it, as the single greatest power, could control. America was also able to further its aims in Europe by taking the burden of intervention from France. French troops could return home, where their presence eased fears about the postwar redevelopment of Germany that had become central to US policy. Economically, the war in Vietnam and the neighbouring states might appear to have been an irrational enterprise for America, costing thousands of lives and billions of dollars. It was much less so for some American-based corporate interests. The costs were borne by the whole of society rather than just those who profited from keeping the region open for access to its markets, cheap labour and sources of raw materials, not to mention the domestic profits made by those involved in the provision of military hardware.[53]

After the war the territory seemed to follow the map. Cold War fiction was in places converted into the fact of a self-fulfilling prophecy. America did achieve a victory of sorts.

Southern Vietnam was left in ruins. The country as a whole was not able to provide an appealing picture of the socialist alternative: 'The battering of the peasant society, particularly the murderous post-Tet accelerated pacification campaigns, virtually destroyed the indigenous resistance by destroying its social base, setting the stage for the northern domination now deplored by Western hypocrites – exactly as had been predicted many years before.'[54] A policy of blocking aid helped to ensure the survival only of harsher elements whose desperate policies were used to justify the initial attack.[55] The massive bombardment of Cambodia in the early 1970s and American support for corrupt puppet regimes were significant factors in the subsequent brutalization and rise to power of the Khmer Rouge. American bombers used maps that were out of date and of too small a scale, failing to show many new settlements which suffered heavy civilian losses as a result.[56] The excesses of the Khmer Rouge provided more circular evidence of what were supposed to be the inherent evils of communism. A similar logic applied when America leapt upon any threatened or actual outbreak of socialism or radical democracy in South and Central America. The states attacked overtly by America or through its proxies were not Soviet clients, as was claimed to justify intervention. Uprisings such as those in Cuba in 1959 or in Nicaragua in 1979 were local and independent rebellions against repressive regimes. They were forced closer to the rival superpower by American intervention more than anything else, having nowhere else to turn for support. The reality shifted under the weight of the map.

Here again we are on a ground that cannot be reduced to map or territory, a global arena in which fictions can be played out to a point at which they become indistinguishable from or create new realities, where the map merges onto the territory. At any particular instant one or other term in the opposition may be identified as holding the balance of power, but the relationship remains complex. Economic factors such as imperial access to raw materials and markets do not exist in isolation. They are part of a broader whole and cannot ultimately be separated out from political or ideological dimensions. In some respects the redrawing of the map attempted by the American state may have been out of proportion to the purely economic importance of the trading relationships at stake, for the country if not for some of the multinational corporations that had most to gain. Political and ideological threads were interwoven with the economic, creating complex cultural fabrics that rest on no single determining ground. The American political system gives considerable influence to corporate entities and elements of the military-industrial complex, each of which have vested interests in overseas intervention. Alternatives might be found for certain of the raw materials or markets lost as a result of 'Third World' rebellions. The prevalence of Cold War ideological mappings led to an exaggeration and generalization of individual difficulties, however, and an overreaction based on global worst-case scenarios: 'A massive edifice of intervention can be built upon a relatively narrow foundation.'[57] Economic imperatives were not defined in isolation from either Cold War or frontier mappings. A potent mix of economic interest and frontier values providing an impetus for overseas intervention exists particularly in the 'Southern Rim' or Sunbelt region, where many defence and related industries were located during and after the Second World War. A remapping of economic and political power has occurred in which this region (itself defined generally by a line on the map, the 37th parallel) has in some respects overtaken the traditional power centres of the Northeast.[58]

A blend of discursive and physical actions also underlay the initial colonization of the New World. Columbus used formal devices to establish the fiction of sovereignty over the islands at which he arrived, a fiction that was to become material fact as his map of the region was applied directly to the territory. After making his first landfall in what are now known as the Bahamas he delivered a proclamation through which he was understood to have taken possession of the land. Understood by his own people, that is. The proclamation was made in Spanish and was unintelligible to the natives even if they had been present, which seems doubtful. According to this legal fiction the indigenous peoples lost their land through their own choice because they did not contradict the Spanish claim. From 1513 this process was formalized in the pages of the *Requerimiento*, a document in Spanish that the conquistadores were supposed to read aloud to all newly encountered peoples. It informed them of their rights and obligations as vassals of the King and Queen of Spain. If they were obedient they would be rewarded. Otherwise their punishment would be harsh and the fault would be their own. The fact that they would not be able to understand any of this was conveniently ignored. Discourses such as these were an integral part of the colonial enterprise. Is this just an ideological cover for the brutal exercise of naked power, Stephen Greenblatt asks. Maybe it is in many cases, he suggests. But we should take care not to naturalize the colonial process: 'The possession of weapons and the will to use them on defenceless people are cultural matters that are intimately bound up with discourse: with the stories a culture tells itself, its conceptions of personal boundary and liability, its whole collective system of rules.'[59] The colonial enterprise became an important part of the rise of capitalism at home, but this was not automatically pre-programmed according to economic dictates alone. Ideological factors were important not just as *post hoc* justifications. They helped to determine the fact that native reality could be treated appropriatively, rather than being respected in its own right.

The same kind of complications arise with the map or grid drawn onto the territory. The Spanish settlements in Latin America that pre-dated many of those established in the north of the continent were laid out in grid-form according to a strict code. Abstractly imposed as they were, these grids may appear to have been merely superstructural in character. But the grid was also an instrument of production, as Henri Lefebvre suggests: 'a superstructure foreign to the original space serves as a political means of introducing a social and economic structure in such a way that it may gain a foothold and indeed establish its "base" in a particular locality.'[60] The imposition of a geometrical urban space in Latin America was 'intimately bound up with a process of extortion and plunder serving the accumulation of wealth in Western Europe; it is almost as though the riches produced were riddled out through the gaps in the grid.'[61]

Where debates within Marxism come unduly to fetishize an isolated economic or material level to which all others are subordinated they might themselves be indulging in the kind of narrow 'rationalism' associated with Western imperialist dynamics. There is a danger, perhaps inevitable, of failing to transcend the terms of the phenomenon under discussion. A rigid distinction between the spheres of a narrowly defined economy and all others is itself a product of capitalism. An analysis that continues to work strictly within these terms will thus remain trapped within capitalism's own logic. In one sense this is precisely what is implied by Marx when he argues that an increasing process of rationalization will eventually tear aside the veil that obscures the exploitative nature of capitalism and lead towards revolution and the creation of a fair and rational society. Capitalism's own dynamic is expected to bring about its downfall. There may be elements of truth in this, but the argument fails to account for the complex nature of cultural cartographies that cannot be reduced to a point of objective transparency.

The potency of such mappings is suggested in Stuart Hall's reading of the Thatcherite political project. Hall utilizes Antonio Gramsci's notion of hegemony to suggest one reason for the longevity of the Thatcher regime in the face of economic decline and spiralling unemployment, an argument that might go some way to explaining a fourth Conservative election victory in a 1992 election conducted from the depths of recession. Under Thatcher, ideological arguments focusing on topics such as race and law and order were used to redefine problems and to shortcircuit anger and arguments that might otherwise have struck at the basis of the system itself. A reactionary common sense was articulated, 'one that has the power to map out the world of problematic social reality in terms of clear and unambiguous moral polarities.'[62] Such grids may be drawn from on high and take little account of existing mappings. This is often the case in imperialist maps such as those imposed in America and Israel. But the process can be more subtle. The Thatcherite map sought to make recognizable connections with the everyday experience of the inhabitants of its territory. Real problems in areas such as housing and employment were fitted into a racist and authoritarian grid that appeared to resonate with experiences of life on the ground.

The claims to objectivity of one map can be challenged by another. Class struggle may be manifested in the struggle for control of the dominant system of cartography, implying as it does the definition of the reality upon which other actions are based. Existing boundaries can be challenged and replaced with new ones. Contour lines can be redrawn. A new mapping can take on the dominant, just as Peters challenged Mercator.

Symbolic strikes might be made, such as the attempt by a character in Joseph Conrad's *The Secret Agent* (1907) to blow up the Greenwich meridian. Maps might also be used in the defence of those they have traditionally oppressed, although they remain potentially double-edged. Thus Hugh Brody expresses the reservations with which he agreed to take part in a land-use and occupancy study of the territory in British Columbia occupied by a Beaver community. He was aware that maps such as those he was drawing had been used before to rob people of their land. But, as one elder argued, the Beaver had to make their presence known if they were to survive. Even if Western notions of mapping and property were essentially alien to their culture, the time had come to insist on their right to be mapped onto the land in a way that would be effective in the world by which they were threatened.[63]

Much debate continues to be conducted in terms of a strict polarity between map and territory. Those who question the distinction are often labelled themselves as idealists, particularly by those on the left who follow the example of Lenin. He refused to accept the possibility of any alternative approach. The question, as he put it in his attack on the followers of Ernest Mach, 'is not of this or that formulation of materialism, but of the antithesis between materialism and idealism, of the difference between the two fundamental lines in philosophy.'[64] An inflexible opposition is imposed by those who prefer the reassurance of a mapping that dismisses the complications I have emphasized. This is not surprising. Where the debate between materialism and idealism is conducted in terms of the politics of the left, otherwise academic questions can become urgent matters of strategy and practice in an area where feelings run high. The result seems to be a series of inversions and counter-inversions. Marx began with a declaration of intent to invert Hegel. His own work has also been inverted, by Marxists, neo-Marxists and non-Marxists seeking to emphasize the influence of elements of the 'superstructure', or by those who advocate a return to Hegelian idealism. These manoeuvres have, in turn, been re-inverted in reassertions of materialism or economism. A perspective able to embrace complex cultural cartographies might better be achieved by an attempt to displace the terms of this debate, to deconstruct the opposition between map and territory.

8 Deconstructing the Map

Deconstruction cannot limit itself or proceed immediately to a neutralization: it must, by means of a double gesture, a double science, practise an overturning of the classical opposition and a general displacement of the system.

Jacques Derrida[1]

Maps such as those imposed by the forces of Western imperialism or patriarchy remain active presences on the territory. They cannot simply be wished away in an immediate act of displacement. An initial moment of inversion is required if existing mappings are to be challenged. The dominant map might simply be turned upside-down, our view of the world defamiliarized and its provisional status recognized. The same kind of operation is required to deconstruct the opposition between map and territory. The map, traditionally relegated to a secondary position, has to be foregrounded before we can go on to displace the terms in which the opposition exists. If we cannot immediately escape existing frameworks an opening movement of inversion can begin to undermine their usual operations and exclusions. It is only in this way 'that deconstruction will provide itself with the means with which to intervene in the field of oppositions that it criticizes, which is also a field of nondiscursive forces.'[2]

This is an important point, ignored in many characterizations of deconstruction. To seek to move immediately to the ultimate stage of displacement, to escape the terms of the opposition, would be to ignore the very real powers and effects that inhere in the systems founded on existing mappings. It would be an attempt to move into a vacuum that would leave such powers and effects untouched, 'and would be to give free reign to the existing forces that effectively and historically dominate the field.'[3] It is in this sense that the moment of inversion is important in the articulation of a Marxist materialism that sets itself against Hegelian and/or bourgeois idealism. An inversion that places the emphasis on the economy can be a powerful assault on capitalist fetishizations of superstructural phenomena such as legal and political institutions. Defenders of existing Western capitalist systems celebrate formal notions of democratic freedom. A counter-emphasis on the economic dimension reveals numerous instances of inequality, exploitation and compulsion that might otherwise be effaced (notwithstanding the spurious claim that capitalism is founded on a 'free-market' economy). To say that superstructural levels are inextricably implicated in the economic is not to deny the possibility of such an argument. The problem comes when such a reassertion of the importance of the economic dimension is elevated into a virtual metaphysics rather than a matter of political practice. The grid applied for a particular purpose becomes frozen and opaque and is imagined to exist on the territory itself.

This remains a serious difficulty and goes a long way to explain the mutual hostility that has often existed between the philosophy of deconstruction and the politics of the left. The two are not always compatible and cannot simply be yoked painlessly together. Capitalism can be seen as an amalgam of map and territory, a complex and multifaceted series of mutually reinforcing structures that maintain its now almost global hegemony. Marxist and other socialist discourse offers an alternative map, a critical analysis from another point of view. Various readings suggest how the territory and its existing mappings might differently be understood or changed. Both critique and practice are based on certain underlying values and ideals. These cannot be proclaimed to be absolute, grounded as they are only in particular cultural cartographies. Marxist theory remains at least partly within the paradigm of Enlightenment rationalism, for instance. We can still stand by the values inscribed on our own maps, however. It would be illusory to pretend to be able to speak from any position exterior to one grid or another. It is possible to combine the notion that our own values are ultimately arbitrary with a passionate belief in them. In the heat of combat, of course, the former is always likely to be sacrificed to the latter.

What about the means to achieve the realization of a particular set of values? If we believe, say, in the proposition that all people have in general an equal right to be well fed, clothed, housed and recognized by their peers, so might some advocates of capitalism. To what can we appeal to settle the matter? The quick answer would be to reel off some facts about global inequality, about world hunger made more acute by the actions of multinationals, about imperialism, domination, environmental destruction and the way people are judged and rewarded as if the good or bad fortune that comes their way were primarily anything to do with their own merit rather than the social fate they inherit. But there are no neutral facts for us that are free of interpretation. Everything that becomes meaningful does so by being fitted into, or challenging, some existing map. We seem at this point to reach an aporia, to use a favourite poststructuralist term, an unresolvable dilemma. We might argue for our own position, equipped with an interpretation of what are considered to be the relevant facts. We might argue it very strongly and act upon it. No more than any opponent, though, could we expect to find any fundamental ground, any unmapped territory on which to base our position. It is tempting to try to sidestep this problem by making a distinction between abstract theoretical analysis and practical politics. But every theory implies a politics and every politics a theory. The two cannot in the end be kept apart. When a pressing tactical or strategic decision has to be taken we nevertheless have to come down on one side or the other, without the luxury always of philosophical nicety. The two cannot always be squared, which is uncomfortable. Politics is an area more for the abuse than the use of philosophical absolutes, tempting though they are.

If Marxist or other socialist theories provide useful maps of capitalism they have also been accused of imperial activities of their own, principally in their claims to sovereignty over territories such as those of sexual and racial oppression. It may be perfectly legitimate to argue that such forms as they exist in a capitalist-dominated world cannot fully be understood without the aid of a class-based analysis. Yet traditional Marxist or socialist discourses alone cannot claim fully to map either domain. Absolutes again have to give way to more proximate matters of strategy and tactic. Situations are bound to arise in which one or another aspect of a group's oppression has to be treated as a priority, when choices have to be made between courses of action that may not be compatible. The issue of inversion and displacement also arises as a vexed strategical point in questions of sexuality. The categories masculine and feminine are cultural constructs, binary opposites imposed in a framework that tends to disallow alternative possibilities. It may at first be necessary to invert the opposed terms, to valorize the figure of 'woman' where man is dominant as a initial step towards challenging the entire map and as a way of intervening in the existing network of forces. This has long been the aim of radical feminists. In some cases woman has in the process been evoked as an eternal essence, celebrated in a way doubtless important in overcoming years of oppression but unhelpful in the longer-term effort to question the basis of the binary opposition on which the term is based.

Patriarchal discourses tend to reduce women to the status of the simulacrum, the work of artifice that stands as 'other' to a privileged reality defined in terms of the masculine subject. The effect may be oppressive, denying any fundamental reality to the existence of women, but it could also offer a way of undermining the entire framework. For Jacques Derrida: 'Woman is but one name for that untruth of truth.'[4] Denied and negated in a whole complex of hierarchical oppositions in which the masculine is associated with the rational, the active and the cultural, and the feminine with the irrational, the passive and nature, 'woman' can be valorized as a figure standing for all those uncharted territories repressed as a condition of the assertion of patriarchal reality.

What is offered here, in what Alice Jardine terms *gynesis*, 'a putting into discourse of "woman"', is 'a map of new spaces yet to be explored'.[5] As Hélène Cixous puts it, the 'dark continent', the repressed landscape of the other, 'is neither dark nor unexplorable. It is still unexplored only because we have been made to believe that it was too dark to be explored.'[6] To a masculinity defined in terms of the search for solid ground and firm foundation, Luce Irigaray contrasts a woman whose existence is fluid and unbounded, turning to her own ends precisely the kind of patriarchal opposition we saw in Chapters 3 and 4. 'Erection is no business of ours,' she declares: 'we are at home on the flatlands.'[7] Men have also in some cases sought to move onto this alternative ground in an effort to escape repressive sexual frameworks. Here we find the 'Wild Men' of men's groups reverting to the Texas wilderness in a search for the kind of space mapped in Nathaniel Hawthorne's *The Scarlet Letter* (1850), a forest or wilderness landscape which allows for the expression of impulses and emotions forbidden by the civilization to which it stands opposed. Borrowings from Native American ritual are also used as an alternative mode of communication, including an equivalent of the 'sweat lodge' purification rite, another attempt to reach a more authentic level of experience. Robert Bly, the principal advocate of the 'Wild Man' movement in the United States, also locates this state of mind in swampish and watery regions rather than on culturally imposed dry land.[8] Excursions into such domains achieve no more than a partial escape from existing frameworks, however. A different and less restrictive notion of maleness may be articulated, but the binary sexual mapping tends to remain intact. Bly, for example, insists on the existence of a 'deep male' buried in the psyche of the modern man. The opposition between so-called 'primitive' and 'civilized' modes is also left largely unchallenged. A darker side can be seen in the survivalist movements of those seeking, often in an implicitly violent, Darwinian manner, to gain the primitivist skills they believe will enable them to survive any cultural apocalypse that might in the future drive humanity back into the woods.

Figuring an uncharted 'dark continent' landscape such as that of the American wilderness in terms of woman is far from new, of course. Precisely such an image formed a guiding metaphor for many early understandings of the supposedly 'virgin' territory. In one sense what appeared to be offered was a regression from the difficult adult concerns of the Old World to the womblike environment of a primal 'feminine' landscape. For Annette Kolodny this was the only way the pastoral dream in America could be sustained after a certain point. She finds the process dramatized in the work of writers such as James Fenimore Cooper by the figure of the frontier hero who can only retain a close relationship with the landscape in a lone existence founded on a rejection of adult sexuality. One way of reading the oscillation between articulations of the wilderness as real or unreal, paradise or hell, is in terms of a tension between such an outlook and a subsequent impulse to act upon and to 'master' this supposed femininity.[9] It is important to have no illusions about the ways in which such sexual figurations might be used if they are not to become just another way of asserting patriarchal domination or any pre-given essence of 'woman'. Derrida himself takes care to set out the two distinct phases required:

> Saying that woman is on the side, so to speak, of undecidability and so on, has only the meaning of a strategical phase. In a given situation, which is ours, which is the European phallogocentric structure, the side of the woman is the side from which you start to dismantle the structure. So you put undecidability and all the other concepts which go with it on the side of femininity, writing and so on. But as soon as you have reached the first stage of deconstruction, then the opposition between women and men stops being pertinent. Then you cannot say that woman is another name, or a good trope for writing, undecidability and so on.[10]

At this point a new, deconstructive reading of the existing map becomes possible in which the arbitrary status of the existing boundaries is apparent. There are problems with this approach. As several critics have pointed out, the grand, almost metaphysical generalization and figuration of the position of 'woman' can itself be an imperialist gesture that fails to account for the specific realities faced by women.[11] We should also beware of any tendency to leap too rapidly into the void, to diagnose prematurely (as Baudrillard does) the achievement of a sexual displacement that remains at best a possibility on the horizon.

Existing frameworks can only effectively be challenged from within, even if some space must be opened up for a redrawing of the mapped territory itself. Thus I have constantly used the term 'cultural', arguing for the importance of cultural cartographies that establish the reality we inhabit, while at the same time questioning the opposing concept of nature without which the former has no meaning. The dilemma is the same as that posed by the term 'literal' to which the metaphorical was counterposed in Chapter 4. As I suggested there, the term 'literal' has a currency and a potency that cannot simply be ignored in an impulsive effort immediately to displace the terms. The metaphorical or the cultural may have to be emphasized, although ultimately considered redundant in its dependence on the term to which it is opposed, if any impact is to be made on the existing debate. Similar problems are encountered with the use of terms such as the real, reality, objectivity and their contraries – the image, representation, fiction, myth, and so on.

The title of my first chapter suggests an inversion rather

than a displacement: the map that *precedes* the territory. The moment of inversion is necessary as an initial response to those who would put the territory first. But the map cannot be treated in isolation. Map and territory become implicated mutually in one another. An adequate topographical metaphor would be rather different from Marx's model of a foundational base upon which sits a superstructure. We might instead look to the irresolvable structures generated in the graphic art of M.C. Escher, where hierarchies are strangely undermined and levels of reality mixed. In *Print Gallery* (1956) we are shown a boy looking at a picture in a gallery. Yet as the eye follows the picture around it is found to include the gallery itself in which it appears to be displayed: 'Maps [. . .] contain the ground that contains them', as a character in Margaret Atwood's *The Robber Bride* (1993) puts it.[12] The reality and the representation are thus inseparable. A hierarchical relationship is replaced by what Douglas Hofstadter terms a 'strange loop', or a recursive structure in which each level appears to be determined or influenced by the others in a dialectical process that cannot ultimately be resolved.[13] The map neither obscures nor is a simple product of the territory; the two are intertwined, as if the map has been woven into the very ground on which it stands. This is not a collapsing together of different elements into a single, undifferentiated unity but a model in which different facets are in some way fused together in mutual implication. It does not rule out the possibility of arguing for specific identifiable determinations in particular contexts such as one phase or another of the colonial process.

Once opened up to the possibility of deconstruction maps can be read in very different ways. They may have played an important role in the achievement of colonial domination, but they can also be used to make visible the process through which domination was gained, and so to oppose it. In general maps tend to be read passively, as charts of a taken-for-granted reality. But they can be reactivated. When read in this way, the place-names on the contemporary map of America:

> reveal lost origins and removed meanings that recount anything but a closed semiotic system of homothetic correspondences. As soon as the map reader questions the map's writing, the microhistories embedded there break out like a cancer

> all over the map's surface, calling into question its very authority. To interrogate the map is to unfix it, to read it kinographically, in which case its static being becomes dynamic process; its towns and cities, kinemes; its boundaries, behaviour.[14]

The entire rhetorical process can be brought back into the foreground and the constructed unity broken down into its component parts. Rather than merely *representing* an abstract and homogeneous space, the map can be seen to assert it, 'the map as a manifestation of the desire for control rather than as an authenticating seal of coherence.'[15] Enlightenment 'rationality' as expressed on the map may be shown for what it really is, an enterprise implicated in a process of violent domination.

Precisely because it is founded on processes of universalizing closure, the map becomes an ideal site for a deconstructive project. What might be revealed, as Graham Huggan puts it, is 'the exemplary structuralist activity involved in the production of the map (the demarcation of boundaries, allocation of points and connection of lines within an enclosed, self-sufficient unit) [that is traced] back to a "point of presence" whose stability cannot be guaranteed.'[16] Gaps and inconsistencies on the map can be highlighted in an attempt to undermine the wider discursive system within which it is embedded. This is the effect of José Rabasa's reading of Mercator's *Atlas*, in which he finds a plurality of semiotic systems and levels of meaning underlying what is presented as an objective totalization of the earth's surface. Information taken from subjective first-hand accounts of travellers is inscribed onto the pages of the *Atlas* in an abbreviated, objectifying form that masks the process through which knowledge is produced. The written section of the work contains a series of taxonomical headings that sift an otherwise unmanageable infinitude of details into a form that can easily be encompassed. No space is left uncharted, myth filling in where other information is lacking. A hierarchical representation of the continents, depicted in allegorical figures around the world map, shows the rest of the world freely offering its riches to Europe in an inversion of the realities of imperialism.[17] But, as Rabasa argues, the erasures attempted by the *Atlas* are far from perfect. The potential exists for a very different reading, what Huggan terms a 'decolonization of the map':

> a dismantling of the self-privileging authority of the West which also suggests that the relations between the 'natural' and the 'imitated' object which inform the procedures of cartographic representation are motivated by the will to power and, further, that these relations ultimately pertain neither to an 'objective' representation nor even to a 'subjective' reconstruction of the 'real' world but rather a play between alternative simulacra which problematizes the easy distinction between object and subject.'[18]

Every map offers only its own perspective on the world, however objective it may appear or claim to be, a perspective that implies a particular assertion of reality. The map-like view of Manhattan from the top of the World Trade Center in New York, for example, might appear to be an all-embracing vision of the city from a height that permits a neutral overview. But the viewer of the map spread out on all sides remains firmly entrenched within a particular perspective, looking out from the summit of a bastion of finance capitalism. The view from the World Trade Center emphasizes the power and rationality of ranks of gleaming skyscrapers while effacing realities of life on the ground such as poverty, homelessness and the city's financial crisis. Alternative mappings might be able to reflect some of these terrestrial factors. The journalist and reformer Jacob Riis of Doctorow's *Ragtime* (1974) makes maps that depict the ethnic populations of Manhattan as 'a crazy quilt of humanity'.[19] Such a mapping might have something to say about ethnic inequalities and the spatial practices through which they are maintained, although the colour-coding adopted is that of dominant convention with racist overtones: black for the African, green for the Irish and yellow for the Chinese. Other kinds of mapping may be able to chart the sources of poverty and disease. The classic historical example is the map produced by Dr John Snow during the London cholera epidemic of 1854. With cases of infection marked onto a street plan, the map provided important evidence to support his belief that the disease came from the water supplied by a communal pump.[20] The social reformer Charles Booth also produced maps depicting the distribution of wealth and poverty in the streets of late nineteenth-century London. Booth's maps clearly had political implications, showing that nearly a third of the city's population lived in poverty. A map based on the census returns of 1981 came up with similar findings.[21] Less accessible to the cartographer are the structures of the financial and computer networks in which finance capitalism is rooted, as Jameson suggested. An exhibition of images of microprocessor circuitry at New York's Museum of Modern Art gave some sense of the intricacy of the structures involved. Some of the works resembled maps of this remote territory, but they could hardly be appealed to for understanding of the economic, political and ideological practices involved in its upkeep. The cyberscapes of the computer Internet are also resistant to other than localized mappings, despite the prevalence of cartographic imagery in many textbooks. Searching mechanisms enable the user to map significant expanses of the network, but we should perhaps be grateful that it retains an essentially anarchic quality despite its origins in the US Department of Defense.

A striking illustration of the illusions of objective mapping is provided by a map of the midtown area published by the Manhattan Map Company and claiming to be 'the most detailed map in the world'. Using an axonometric projection this map gives an extraordinarily detailed three-dimensional view of midtown Manhattan which at first appears to be the ultimate development of the tradition of Renaissance perspective views discussed in Chapter 3. But this map goes a stage further in its claim to objective representation. It is not a linear perspective view at all, but an isometric map that reproduces each building on the same scale. The view it presents is an impossible one. There is no diminution in the size of buildings as the map recedes and the vanishing point is infinity. The cartographers have combined the best features of a direct overhead view, showing the abstract map-layout of streets, and that taken from a 45-degree angle that picks out the elevations of buildings otherwise flattened and made unrecognizable. The result is a map that depicts minute detail and is in one sense highly accurate but that is entirely out of keeping with the plastic experience of life in Manhattan. A bustling, crowded, noisy and poverty-ridden cityscape that generates feelings of claustrophobia in many of its inhabitants is rendered into an abstract pattern that gives a strong impression of light, space and cleanliness. The absence of the particular linear form of recession that has come to be suggested by the term 'perspective' is itself a very particular and persuasive perspective.

The map of Midtown Manhattan occupies a place in a historical sequence in which a gradually increasing impression of objectivity has been created in the development of cartographic techniques, some of which have already been discussed. From the imagined bird's-eye views of Renaissance artists to the use of balloons, aircraft and satellites, a higher and higher viewpoint has been gained. The first spacecraft to leave the orbit of the Earth provided what appeared to be the first truly objective view of the planet from beyond its physical confines. Even this view cannot escape implication in the territory, however. The image of the globe, which seems so objective a representation of the Earth when compared with the distortions caused when it is projected onto a flat map, has long been used for its rhetorical powers. The globe is depicted in the hands of a monarch or a god to symbolize a sovereign rule. The image of the planet from space is used today by those wishing to emphasize the fragility and vulnerability of the planet. To whichever end it is put the image is a potent one used in the construction of specific terrestrial meanings.

The view from above has always been implicated in questions of military power, from the first wartime use of observation balloons in the late eighteenth century to the mapping of destruction in the Gulf war. It is a view that can easily be distorted for political ends. Satellite pictures were used by America to justify its intervention in the Gulf in 1990 on the basis of an Iraqi threat to Saudi Arabia. They were almost certainly doctored to exaggerate the threat, a practice employed by the US during the earlier war between Iran and Iraq.[22] Satellite images are authoritative and convincing, although not easily read by the non-expert. They do much, for example, to increase our faith in the predictions of weather forecasters. All forms of remote sensing entail the manipulation of data, most of which is converted into visual mappings that can be processed in a variety of ways. Images are 'enhanced' (a suggestive term) and their meanings may be altered, even when no deliberate deception is intended. In most cases the images are digital and can be subjected to the kind of computer manipulations described in Chapter 1. Common practices include the sharpening of boundary edges or the enhancement of images through the use of filters and screens: processes analogous to the imposition of other cultural grids through which preferred meanings are constructed.[23]

Under the gaze of modern state-of-the-art satellite and computer mapping techniques it seems that no physical feature of the world can remain hidden for long. The world appears to have become a giant panopticon. Satellite images have revealed previously unknown lakes in Canada while an islet off the Atlantic coast is named Landsat Island after its space-age discoverer. A map of nothing less than the entire universe and its origins was produced in April 1992 by NASA scientists using data from the Cosmic Background Explorer satellite, Cobe. It appeared to explain why matter had taken the form of galactic structures suspended in a void (perhaps the ultimate in mappable territories) rather than being distributed evenly according to the dictates of entropy. For those who wished to measure authenticity in terms of access to originary pre-structures, Cobe seemed to have the last word, mapping traces of primordial ripples at the edge of the cosmos.

The use of radar, microwaves, infrared and ultraviolet light has combined with computer technology to provide instant maps of almost anything, from mineral deposits and geological fault-lines to the hole in the ozone layer and the floors of the oceans.[24] Again these are perspectives inextricably linked with relations of power and exploitation. They create particular understandings and can bring about a variety of changes in the territory itself. Satellite mappings of global crop developments might be used to help poor countries make the most of their resources, but are more often in the possession of those who can use them to make lucrative decisions either in their own agribusiness enterprises or the commodities futures market. The uranium deposits underlying the Black Hills of South Dakota were located by satellite surveillance in the early 1970s, a development that might explain the level of force that was used against radical Native American groups whose insistence on the rights granted in earlier treaties stood in the way of corporate mining interests.[25] The previously uncharted wilderness of the Amazon basin, the rainforest that seemed to defy Western cultural grids, was mapped from the air by radar in the 1970s.[26] The declared aim was to aid settlement by landless peasants, but the effect has been deforestation and poverty for many of the settlers brought by the new mappings to land that often could not sustain them. A number of new projects are under way to map the rainforests on a global basis.[27] The intention is to chart the threat to a vital environmental resource, but mappings such as these can always have unforseen consequences and can be used for the achievement of other ends. Satellite technology offers the possibility of mapping vast areas of the globe that have only been charted poorly or approximately. It also threatens to open such areas to a variety of interventions. Maps have been made of the previously unseen landscape beneath the ice masses of Antarctica, opening the way to the exploitation of one of the few untouched wilderness territories on the planet. At the other end of the scale, the project to map the structure of the human genome promises a new understanding of our genetic makeup. Many benefits may result, but difficult ethical questions will be asked about the use of knowledge of the fate mapped out by genetics, not to mention the clash between those who favour cultural rather than biological explanations of human activity.

The most advanced military satellites have gone beyond mere surveillance to the compilation of digital maps from which databases of the earth's surface are being produced by the US Defense Mapping Aerospace Agency. These can be used for training exercises, simulated rehearsals or actual missions. Computer scientists at NASA's Jet Propulsion Laboratory at Pasadena have created a three-dimensional simulated map of the world. The fighter-bomber pilot of the near future will be equipped with a virtual reality helmet in which such maps of the territory under attack can be projected onto his field of vision. The Tomahawk cruise missile already uses this kind of technology, as was seen amid much triumphalist publicity during the war in the Gulf in 1991:

> For most of its lethal journey, the Tomahawk navigates through a radar altimeter which compares the topography of its flight path against detailed computer maps stored in its memory. As it reaches the 'terminal end point', a new guidance system takes over, with a small digital camera comparing the view from the nosecone against a library of stored images prepared from earlier satellite photography of specified targets.[28]

The abstracted, techno-mapping view from above is fallible, however. The war in the Gulf may have been represented as its ultimate achievement, but it remains blind to many of the more complex socio-cultural or political realities on the ground. Secret U-2 flights over military installations in the Soviet Union from 1956 provided President Eisenhower with objective evidence of overwhelming American missile superiority. But this was of little use against those who claimed that the United States was being overtaken or even betrayed. The combination of Cold War ideology and economic and political interest was more potent, and would probably have remained so even if the president had been able to show publicly the covert view from on high. Even in Iraq, despite the pounding his forces took and the loss of Kuwait, Saddam Hussein was not removed from power. Fake tanks and missile launchers were able in some cases to fool those who looked down from above, a tactic also employed by the Vietnamese. The lesson of Vietnam remains. However destructive they might be, the reality-claims of the abstract imperial grid are limited. If we wanted a symbolic alternative to the overhead perspective of the digital terrain mapper we might find it in the networks of underground tunnels within which the Vietnamese were able to maintain their resistance out of sight of the panoptic gaze. Here we might encounter a very different cultural cartography, akin more to the rhizome of Deleuze and Guattari than to the abstract map imposed from on high.

The view from above need not be proclaimed as objective or objectifying. It was not for the Russian artist Kasimir Malevich. His avowedly non-objective art was the product of a technological environment the ideal image of which was the aerial view, the new perspective opened up by the advent of the aeroplane. In the ascent to the heights of non-objective art, he writes, 'The familiar recedes ever further and further into the background . . . The contours of the objective world fade more and more and so it goes, step by step, until finally the world – "everything we loved and by which we have lived" – becomes lost to sight.'[29] When Gertrude Stein flew across America she saw 'the lines of cubism' on the ground rather than an objective mapping of the landscape.[30] We might also begin to see the outlines of other cultural forms: the shadow of old field systems and settlements, different models of existence, or features such as military establishments that have been erased from the map.

If Western imperialist claims of access to absolute reality looked for their authority to Newtonian science and its forerunners in the art and science of linear perspective, rival bids have been made in recent years in the name of theories of relativity (Einstein), uncertainty (Heisenberg), undecidability (Gödel) or chaos. The possibility of a complete mapping is thrown into doubt, for example, in Benoit Mandelbrot's fractal geometry, a study of those awkward, uneven forms that fail to conform to the world of Euclid. What might appear to be clearly bounded coastal territories prove elusively unmappable when an attempt is made to chart the finer detail: 'When a bay or peninsula noticed on a map scaled to 1/100,000 is reexamined on a map at 1/10,000 subbays and subpeninsulas become visible. On a 1/1,000 scale map, sub-subbays and sub-subpeninsulas appear and so forth. Each adds to the measured length.'[31] Similar difficulties may be encountered in the mapping of national borders. Where they are defined by means other than the geometric grids of latitude and longitude their precise length and detail might resist any definitive measurement.[32]

How far these sciences of uncertainty can usefully be mapped onto the cultural sphere remains questionable. Once more there is a danger of inversion taking precedence over displacement, of the assertion of patterns of randomness and chaos as if they were an answer to the confident certainties either of classical Newtonian science or the uses to which it has been put in support of imperial domination. To make such an assertion, as some have done in the name of the postmodern, is to risk at least two errors. First, the sciences of chaos do not in fact present an utterly chaotic perspective. What is detected is another kind of order arising from within chaos, a regularity in the occurrence of irregularity such as the whirling fractal pattern that is replicated again and again at different levels of magnification. Fractal geometry is not presented by its founder as an unknowable quantity. Quite the opposite. What Mandelbrot proposes is an extension of the boundaries of geometric mapping into new territories: 'Scientists will (I am sure) be surprised and delighted to find that not a few shapes they had to call *grainy, hydralike, in between, pimply, pocky, ramified, seaweedy, strange, tangled, tortuous, wiggly, wispy, wrinkled,* and the like, can henceforth be approached in a rigorous and vigorous quantitative fashion.'[33] It is, in other words, nothing less than a geometry of the 'wild', the wilderness, the territory previously unchartable in its own terms and subject only to abandonment, destruction or the imposition of abstract grids from on high. It is another, albeit more subtle, abstraction of the quantitative from a multidimensional reality. On the map of America, an equivalent at the macroscopic scale might be found in the wiggly, spaghetti-like geography of the Beverly Hills area of Los Angeles, a stark contrast to the strict gridiron pattern of the rest of the city. Here the irregular landscape has been encompassed by mappings, both conceptual and cartographical. Physically, Beverly Hills may be a confusing space, but its name offers instant orientation in the cartography of wealth and status. It has also been mapped extensively for the tourist, numerous representations of its winding topography being on sale to chart the addresses past and present of Hollywood stars.

A theorization of the chaotic or the uncertain that remains 'rigorous', 'vigorous' and 'quantitative', or one that provides the basis for a tourist mapping, seems to suggest once more the difficulty, if not impossibility, of extracting ourselves from the dominant cultural maps within which we live. A critique of Western 'rationality' such as that attempted in the previous chapter tends to employ terms and methods of argument implicated in that very same discourse. My use of Native American cartography as an alternative to that of Western imperialism was an inversion of the terms of an argument that had previously been put to precisely the opposite use: to celebrate the culture of the West through the negative reference provided by the Native American other. My intention has been to undermine the terms in which the opposition is put in either reading, but it is hard to avoid falling back into such oppositions. While questioning the all-encompassing mappings of Western imperialism, I have imposed my own mapping onto the wide range of material assembled here. I have stood as if above the territory, mapping onto it the framework through which the argument of the text has been articulated. In this sense *Mapping Reality* embodies its own argument about the imperative to create meaning by imposing maps onto the world. The impression might be given that almost everything can be fitted into the grid of a kind of meta-mapping, a mapping of a variety of other mappings and taxonomies. Is this really the case in any external sense, or is it because the map metaphor engenders a territory of its own that cannot but confirm the images it projects? A similar problem confronts Fredric Jameson. The figure of the map, he suggests, may be too familiar to encompass the matrix of the world behind the computer screens of multinational capital: 'Cognitive mapping, which was meant to have a kind of oxymoronic value and to transcend the limits of mapping altogether, is, as a concept, drawn back by the force of gravity to the black hole of the map itself (one of the most powerful of all human conceptual instruments) and therein cancels out its own impossible originality.'[34]

To admit such an implication in the mappings we impose is to concede the limitations of our own discursive structures, our inability to escape their confines even when they seem oppressive. Yet it is also, in another sense, a disavowal, and each further subclause (such as this) a further disavowal, *ad infinitum*, in a spiral of self-implicatory gestures that in fact create the distancing effect of a still greater claim to an appropriative reality. Where any other kind of register is attempted it is liable to be met with the kind of criticism made of Baudrillard's tendency to lapse into an opaque and at times almost impenetrable rhetoric. Baudrillard's work sometimes takes on a materiality of its own, resistant to the appearance of transparency that so often serves to mask the productions through which meaning is generated. The danger with such forms of discourse is that they risk falling into incomprehensibility, silence or an inability to intervene in important debates.

However resistant to ultimate resolution they might be, the fractal dimensions explored by Mandelbrot remain subject to some kinds of mapping, even if unconventional ones such as those figured in models such as the Koch curve or the 'squig' offered by Mandelbrot himself. Even at this level, and with the admission of limitations, the cartographies involved are not essentially different from the most crude *mappa mundi* or the 'rational' homogeneous space of the Ptolemaic grid: mappings within which some kind of orientation or control of space can be achieved. We cannot escape from these mappings into any entirely uncharted territory, which brings me to the second objection to any too eager an assertion of the chaotic model.

Cultural cartographies can be seen as grids imposed on what might otherwise threaten to become a chaotic void. This is the kind of world charted in the writings of Samuel Beckett, in which the void is just about kept at bay by a desperate maintenance of mappings, games and rituals. These may be fragile structures, but at the level of everyday existence they are more often experienced as solid realities. The once-universal view of the classic Newtonian scientist may be giving way to a conception of the world as complex and plural. This shift in scientific paradigms cannot simply be mapped onto the social and economic landscape, however. It is one thing for images taken from Newtonian physics to be used in part to justify the imperialist project of the West; it is quite another to imagine that a rival scientific discourse can have a similar power seriously to challenge it.

If assertions of chaos offer little hope of countering the potency of existing socio-politico-economic frameworks, what use are they to the everyday realities lived within cultural mappings? All cultural groups impose some kind of cartographic meaning onto the world. Rather than challenging the claim to absolute reality of any one map with an advocation of chaos or cultural void, we should seek change through its displacement by an alternative grid. If Irigaray's image of a fluidity working to dissolve the artificial solidities of a fortress-like construction of the male offers a vivid figuration of the first phase of the process of cultural change, it may be found wanting in the second movement in which some new kind of ground has to be formulated.

It is doubtful that we can have more than an occasional, proximate or abstractly theoretical consciousness of the provisionality of the mappings we inhabit. At moments of upheaval such insights may be possible, but they may also be terrifying, disabling or both. In such circumstances there is the possibility of beneficial change, the redrawing of the map in a way that reduces inequality and oppression. There is also a danger of the imposition of a rigidly reified mapping that treats any questioning of its absolute lines of demarcation as a capital offence. This may be a reversion to the previous grid or its replacement by another of equally dubious merit, something of the alternative currently facing elements of the former Soviet Union caught between the possibility of a reassertion of the old order or a leap into the so-called 'market' economy of Western capitalism. Even when we are aware that the lines of our existing maps are arbitrary and appropriative we cannot simply escape their matrix. That we might consciously be able to disavow and argue against what is mapped out by capitalist, imperialist, racist or patriarchal grids does not mean we can free ourselves from their grasp. Cultural cartographies operate at a level of collective consciousness or unconsciousness and institutionalized social, economic and political practice to which access is not so easily gained. Change is often difficult to achieve. But this might also hold out the possibility of establishing new and more equitable mappings, cartographies that in time could gain a similar hold, providing effective means of orientation but without necessarily becoming entirely reified. Any such mapping would remain a cultural construct. It would continue to precede the 'territory', whatever that really is. Beyond simple oppositions between fact and fiction, map and territory, it would still in a sense be a fiction or map lived as reality or the territory itself. If reimpositions of domination or fundamental inequality are to be avoided, it is essential that the ultimate provisionality of the map remains accessible, if only in part. The map's claim to chart a pre-given territory must be open to challenge and, where necessary, further change: additional drawing and redrawing of the map.

Notes

CHAPTER 1: THE MAP THAT PRECEDES THE TERRITORY

1. *Simulations*, 2.
2. *Lake Wobegone Days*, 91–2.
3. 'Do You Love Me?', 46.
4. *Sexing the Cherry*, 81.
5. Robin Mead, 'Treason on a big scale', 33.
6. *Sylvie and Bruno Concluded*, 169.
7. 'Of Exactitude in Science', 131.
8. *Simulations*, 2.
9. Ibid, 3.
10. Scott Lash, *Sociology of Postmodernism.*
11. *Simulations*, 146; original emphasis.
12. Ibid, 147.
13. *America*, 37.
14. E. Anne Kaplan, *Rocking Around the Clock*, 28.
15. Howard Rheingold, *Virtual Reality*, 100.
16. 'The Reality Gulf', *The Guardian*, 1 January 1991, a translated version of 'La guerre du golfe n'aura pas lieu'; see also the three original essays in *La guerre du Golfe n'a pas eu lieu.*
17. Steven Bode and Paul Wombell, 'In a new light', 3.
18. Fred Ritchin, 'The end of photography as we have known it', 10.
19. Ibid, 12.
20. Kroker, Kroker and Cook, *Panic Encyclopedia*, 14.
21. *America*, 10, 29.
22. *Postmodern Geographies*, 222–223.
23. Kevin Lynch, *The Image of the City*, 43.
24. Soja, *Postmodern Geographies*, 245.
25. Mike Davis, *City of Quartz*, 188.
26. 'The cultural logic of late capitalism', 44.
27. Davis, *City of Quartz*, 252–253.
28. *Late Capitalism.*
29. 'The cultural logic of late capitalism', 52.
30. Ibid, 52.
31. 'Cognitive Mapping', 351.
32. Ibid, 353.
33. 'The cultural logic of late capitalism', 66.
34. Rheingold, *Virtual Reality*, 387.

CHAPTER 2: WORLD VIEWS

1. *The Oxford English Dictionary*, 2nd edition.
2. Erwin Raisz, *General Cartography*, 61.

3. Peter Gould and Rodney White, *Mental Maps.*
4. See, for example, R. Jolliffe, 'An information theory approach to cartography', or A.G. Hodgkiss, *Understanding Maps,* 21–22.
5. Phillip Muehrcke, 'Map reading and abuse', 17.
6. Speier, 'Magic Geography', 311.
7. S.W. Boggs, 'Cartohypnosis', 469.
8. W.W. Jervis, *The World in Maps,* 157.
9. 'Magic Geography', 310.
10. Reproduced in Speier, 'Magic Geography', 319.
11. 'Magic Geography', 321.
12. K.A. Sinnhuber, 'The representation of disputed political boundaries in general atlases', 22.
13. Ibid, 25.
14. *The New State of the World Atlas,* 10–11.
15. J.B. Harley, 'Meaning and ambiguity in Tudor cartography', 22.
16. J.R. Hale, *Renaissance Europe 1480–1520,* 51–52.
17. Harley, 'Meaning and ambiguity', 24.
18. Ronald Rees, 'Historical links between cartography and art', 61.
19. A reading suggested in Harley, following a schema suggested in Erwin Panofsky, *Meaning in the Visual Arts.*
20. Harley, 'Meaning and Ambiguity', 31.
21. David Smith, *Maps and Plans for the Local Historian and Collector,* 91.
22. Ibid, 98.
23. Yolande O'Donaghue, *William Roy (1726–1790): Pioneer of the Ordnance Survey.*
24. A. Philip Muntz, 'Union Mapping in the American Civil War'.
25. Yolande Hodson, 'Fast Maps', in Peter Barber and Christopher Board, *Tales From the Map Room.*
26. Mark Monmonier, *How to Lie with Maps,* 45.
27. Ibid, 88.
28. R. Helgerson, 'The Land Speaks: Cartography, Chorography, and Subversion in Renaissance England', 81.
29. Monmonier, *How to Lie with Maps,* 88–89.
30. See, for example, the map in *The Observer,* 11 April 1982.
31. Seymour Schwartz and Ralph Ehrenberg, *The Mapping of America,* 16; plate 1, 18.
32. Ibid, plate 2, 19.
33. Ibid, 140; plate 84, 146.
34. Ibid, 142.
35. Ibid, 160; plate 96, 164.
36. *Lake Wobegone Days,* 94.
37. Mary Hamer, 'Putting Ireland on the Map'.
38. Norman Thrower, *Maps and Man,* 10.
39. C. Raymond Beazley, *The Dawn of Modern Geography,* volume I, 271.
40. M.J. Blakemore and J.B. Harley, 'Concepts in the History of Cartography'.
41. Juergen Schulz, 'Jacopo de Barbari's View of Venice: Map Making, City Views, and Moralized Geography Before the Year 1500', 448.
42. Beazley, *The Dawn of Modern Geography,* volume I, 247; Pliny the Elder, *Natural History,* 75–79.

43. For illustrations of the first three of these maps, see Rodney W. Shirley, *The Mapping of the World*, 154–155, 158, 165; for the Netherlands map, see Barber and Board, *Tales from the Map Room*, 78–79.
44. Harley, 'Meaning and ambiguity', 34.
45. Schulz, 'Jacopo de Barbari's View of Venice', 449.
46. Ibid, 453.
47. Denis Cosgrove, *Social Formation and Symbolic Landscape*, 110.
48. Schulz, 'Jacopo de Barbari's View of Venice', 468.
49. Cosgrove, *Social Formation and Symbolic Landscape*, 111.
50. Ibid, 113.
51. The three maps discussed here are reproduced in Shirley, *The Mapping of the World*, 13, 168, 218; on the Hondius map, see also Peter Barber, 'The Map Takes On An Ideological Hue', in Barber and Board, *Tales from the Map Room*, 24–25.
52. Harley, 'Meaning and ambiguity', 36.
53. R. Helgerson, 'The Land Speaks', 56.
54. Ibid, 56.
55. Rees, 'Historical links between cartography and art', 64.
56. Terry Hardaker, 'Introduction' to the *Peters Atlas of the World*, 6.
57. Monmonier, *How to Lie with Maps*, 97.
58. Foreword to the *Peters Atlas of the World*, 3.

CHAPTER 3: MAPS OF MEANING

1. *The Order of Things*, xx.
2. Peter Jackson, *Maps of Meaning*, 2.
3. *Culture and Communication*, 51.
4. Ibid, 54.
5. *Structural Anthropology*, 21.
6. *The Sense of Order*, 1.
7. For the classic expression of this approach see Benjamin Lee Whorf, *Language, Thought and Reality*.
8. Edward Sapir, *Culture, Language and Personality*.
9. Ibid, 69.
10. See Arnold van Gennep, *The Rites of Passage*.
11. See Mary Douglas, *Purity and Danger*, 102; E.E. Evans-Pritchard, *Witchcraft, Oracles, and Magic among the Azande*; Clifford Geertz, 'Common Sense as a Cultural System'.
12. *Outline of a Theory of Practice*, 2.
13. Ibid, 37–38.
14. Such as Fredric Jameson in *The Prison-House of Language*.
15. David Harvey, *The Condition of Postmodernity*, 243.
16. Samuel Edgerton, *The Renaissance Rediscovery of Linear Perspective*, 9–10.
17. Ibid, 90; Henri Lefebvre, *The Production of Space*.
18. Edgerton, *The Renaissance Rediscovery of Linear Perspective*, 114–115.
19. Ibid, 162.
20. Rosalind Krauss, 'The Originality of the Avant-Garde', 160.

21. Ibid, 161.
22. Ibid, 161.
23. 'On the Formalistic Character of the Theory of Realism', 81.
24. Jonathan Rutherford, 'Who's That Man?', 23.
25. Ibid, 50.
26. Benedict Anderson, *Imagined Communities*, 19.
27. Muhammad Rumaihi, *Beyond Oil*, 56.
28. For a wonderful spoof, see 'Huge space-filling graphic goes in', *Private Eye*, 1 March 1991, 17.
29. Mark Monmonier, *How to Lie with Maps*, 94.
30. Reuters, 30 April 1992.
31. Edward Said, *The Question of Palestine*.
32. Mary Douglas, *Purity and Danger*.
33. See, for example, Charles Jenks, 'Stone, Paper, Scissors'.
34. Noam Chomsky, 'On US Gulf Policy'; 'The New World Order'; *Deterring Democracy*.
35. *The Vision Thing*, Channel 4, 2 August 1994.
36. Branka Magas, *The Destruction of Yugoslavia*.
37. For example, see the version of the Vance-Owen map reproduced in *The Times*, 4 March 1993; a map compiled from Bosnian sources in *The Times*, 5 January 1994; and a map depicting the ethnic composition of the former Yugoslavia in *The Times*, 6 March 1993.
38. Misha Glenny, *The Fall of Yugoslavia*, 212.
39. Neal Ascherson, 'The new Europe', *The Independent on Sunday*, Review, 9 February 1992, 31.
40. Ibid, 31.

CHAPTER 4: MAPPING THE VOID

1. 'On Poetry', lines 177–180.
2. *The Communist Manifesto*, 83.
3. *Anti-Oedipus*.
4. Robert Lifton, *Boundaries*, xi.
5. See Edward Said, *Covering Islam*.
6. This position is qualified by Deleuze and Guattari in the later essay 'On the Line'.
7. 'Capitalism, Modernism and Postmodernism', 132.
8. *America*, 63.
9. *Waterland*, 11.
10. *A Poetics of Postmodernism*, 43.
11. *Waterland*, 35.
12. Ibid, 2, 15.
13. *The Road to Botany Bay*, 197.
14. Ibid, 198.
15. Ibid, 47.
16. Ibid, 51.
17. Ibid, 68.

18. William Stearn, 'Notes on Linnaeus's "Genera Plantarum"', xiii–xiv.
19. Michel Foucault, *The Order of Things*, 134; *Discipline and Punish*, 148.
20. For examples, see Percy Adams, *Travellers and Travel Liars 1660–1800*, 69.
21. Ibid, 46.
22. John Allen, 'Lands of Myth, Waters of Wonder: The Place of the Imagination in the History of Geographical Exploration', 53.
23. For reproductions of these maps, see Shirley, *The Mapping of the World*, 87; plates 67 and 86.
24. Ibid, 159; reproduction of Gilbert map, 160.
25. Allen, 'Lands of Myth', 55.
26. Carter, *The Road to Botany Bay*, 135.
27. Barber and Board, *Tales from the Map Room*, 82–83.
28. *Of Plymouth Plantation*, extracted in Perry Miller (ed.), *The American Puritans: Their Prose and Poetry*, 17.
29. John Stilgoe, *Common Landscape of America, 1580–1845*, 26.
30. Kirkpatrick Sale, *The Conquest of Paradise*, 79.
31. Leo Marx, *The Machine in the Garden*, 38.
32. Alexander Wilson, *The Culture of Nature*, 97.
33. Dan Stanislawski, 'The origin and spread of the grid-pattern town'.
34. John Reps, *Town Planning in Frontier America*, 246.
35. Boorstin, *The National Experience*, 244.
36. See Hildegard Binder Johnson, *Order Upon the Land*.
37. Boorstin, *The National Experience*, 242.
38. Johnson, *Order Upon the Land*, 198–199.
39. *Culture and Communication*, 33.
40. *Rainforest*, 167.
41. Ibid, 25.
42. *Natural Symbols*, 36.
43. *The Forest People*, 117.
44. Stephen Kern, *The Culture of Time and Space*, 185.
45. See, for example, Arthur Kroker and David Cook, *The Postmodern Scene*, or Arthur Kroker, Marilouise Kroker and David Cook, *Panic Encyclopedia*.
46. *Rainforest*, 79.
47. *The Road to Botany Bay*, 157.
48. Ronald Berndt and Catherine Berndt, *The Speaking Land*, 6.
49. See Yi-Fu Tuan, *Topophilia*, 87–91.
50. *Waterland*, 15.
51. Ibid, 53–54.
52. Ibid, 52.
53. Carter, *The Road to Botany Bay*, 326.
54. 'False Documents', 26.
55. Hayden White, 'The Historical Text as Literary Artifact', 52.
56. Samuel and Thompson, *The Myths We Live By*, 14.
57. Ibid, 6.
58. *The Road to Botany Bay*, 313; my emphasis.
59. See George Lakoff and Mark Johnson, *Metaphors We Live By*, 3; for examples of the use of metaphor in science see Gillian Beer, *Darwin's Plots*.

60. 'On the Line', 26; original emphasis.
61. Ibid, 26.
62. 'On Truth and Falsity in their Ultramoral sense', 180.
63. 'White Mythology: Metaphor in the text of philosophy', 219–220.
64. Ibid, 213.
65. David Cooper, *Metaphor*, 264.
66. *The New Science of Giambattista Vico*, 154.
67. Paul Ricoeur, *The Rule of Metaphor*, 198.
68. *Metaphors We Live By*, 145–146.

CHAPTER 5: THE THICKET OF UNREALITY

1. *The Image: A Guide to Pseudo-Events in America*, 3.
2. *Rise to Globalism*, 193.
3. *The Hidden Persuaders*, 45.
4. *The Society of the Spectacle*, s.47.
5. Ibid, s.6.
6. Ibid, s.10.
7. Ibid, s.31.
8. *The Image*, 37.
9. Ibid, 240.
10. *The Hidden Persuaders*, 13.
11. Colin Hughes and Patrick Wintour, *Labour Rebuilt*, 62.
12. Victoria McKee, 'The ultimate stay at home holiday – With computer technology, travellers can avoid the nasty bits of "abroad"', *The Guardian*, Travel, 16 February 1991, 31.
13. Dean MacCannell, *The Tourist: A New Theory of the Leisure Class*; Erving Goffman, *Frame Analysis*.
14. Henri Lefebvre, *The Production of Space*, 84.
15. 'The Work of Art in the Age of Mechanical Reproduction'.
16. See Janet Wolff, *The Social Production of Art*.
17. Quoted in Dick Hebdige, *Hiding in the Light*, 50.
18. Ibid, 49.
19. See, for example, Claude Lévi-Strauss, *Totemism*; Roland Barthes, *Mythologies*; Baudrillard, 'Consumer Society'.
20. See, for example, Baudrillard, *For a Critique of the Political Economy of the Sign*; Thorstein Veblen, *Theory of the Leisure Class*.
21. *The Americans: The Democratic Experience*, 135.
22. *Simulations*, 25.
23. Ibid, 11.
24. Ibid, 12.
25. 'Travels in Hyperreality', 4.
26. Ibid, 7.
27. Miles Orvell, *The Real Thing*.
28. Eric Hobsbawm and Terence Ranger (eds), *The Invention of Tradition*.
29. *The Enchanted Glass*, 10.
30. Ibid, 33.

31. Ibid, 91–92.
32. Ibid, 171.
33. Ibid, 106.
34. Ibid, 348.
35. Ibid, 123.
36. *The Image*, 57.
37. *The Right Stuff*, 30.
38. Michael Smith, 'Selling the Moon', 184–187.
39. *The Right Stuff*, 215.
40. Ibid, 219.
41. Ibid.
42. Ibid, 266.
43. Walter McDougal, *The Heavens and the Earth: A Political History of the Space Age*, 7.
44. Ibid, 145.
45. 'Leading on Earth and in Space', *International Affairs*, September 1962, 4.
46. McDougal, *The Heavens and the Earth*, 64.
47. Ibid, 174, 178.
48. John Lewis Gaddis, *Strategies of Containment*, 92.
49. John Logsdon, *The Decision to Go to the Moon: Project Apollo and the National Interest*, 93.
50. Quoted in Logsdon, *The Decision to Go to the Moon*, 125–126.
51. 'Special Message to the Congress on Urgent National Needs', 403.
52. Daniel Yergin, *Shattered Peace: The Origins of the Cold War and the National Security State*.
53. See Noam Chomsky, *Deterring Democracy*.
54. Erwin Raisz, *General Cartography*, vi.
55. A.G. Hodgkiss, *Understanding Maps*, 18.
56. John Ager, 'Maps and Propaganda', 5.
57. Yergin, *Shattered Peace*, 211.
58. Alan Henrikson, 'Maps, Globes, and the "Cold War"', 446.
59. Ibid, 448.
60. Ager, 'Maps and Propaganda', 8.
61. See, for example, 'Soviet Distorted Recent Maps; Security Is Believed the Reason', *The New York Times*, 18 January 1970, section 1, 2.
62. Mark Monmonier, *How to Lie with Maps*, 50–51.
63. *The Heavens and the Earth*, 305.
64. Ibid, 304–305.

CHAPTER 6: MAPPING THE FRONTIER

1. *The Road to Botany Bay*, 158.
2. 'The Significance of the Frontier In American History', 38.
3. Ibid.
4. Ibid, 42.
5. Francis Jennings, *The Invasion of America*, 30.

6. William Cronon, *Nature's Metropolis.*
7. In Schwartz and Ehrenberg, *The Mapping of America,* plate 21, 52.
8. Ibid, 148.
9. In Cumming, Skelton and Quinn, *The Discovery of America,* plate 30, 67.
10. Schwartz and Ehrenberg, *The Mapping of America,* plate 50, 95.
11. This is also a feature of other maps in the collection, however, even where no removal of the indigenous peoples has occurred.
12. Louis De Vorsey, 'Amerindian contributions to the mapping of North America'.
13. Simon Berthon and Andrew Robinson, *The Shape of the World,* 160–161.
14. Schwartz and Ehrenberg, *The Mapping of America,* 95.
15. Ibid, 96.
16. Ibid, 119; plate 67, 122.
17. *The National Experience,* 219.
18. Howard Mumford Jones, *O Strange New World.*
19. Purchas, quoted in Greenblatt, *Marvellous Possessions,* 10.
20. Greenblatt, *Marvellous Possessions,* 74.
21. Hernan Cortés, *Letters from Mexico,* 257.
22. *The Journal of Christopher Columbus,* 30, 40, 92.
23. *Marvellous Possessions,* 7.
24. Ibid, 30.
25. Quoted by Hernando Columbus, *The Life of the Admiral by His Son,* 89.
26. 'Narrative of the third voyage of Christopher Columbus to the Indies, in which he discovered the mainland, dispatched to the sovereigns from the island of Hispaniola', 218.
27. Ibid, 222.
28. Ibid, 224.
29. Greenblatt, *Marvellous Possessions,* 78–79.
30. William Phillips and Carla Phillips, *The Worlds of Christopher Columbus,* 169.
31. Howard Mumford Jones, *O Strange New World,* 16.
32. Boorstin, *The National Experience,* 290.
33. *Lake Wobegone Days,* 41.
34. Cronon, *Nature's Metropolis,* 33.
35. Slotkin, *Regeneration Through Violence.*
36. Ibid, 161.
37. *The Discovery, Settlement and present state of Kentucke,* 5; Slotkin, *Regeneration Through Violence.*
38. *Regeneration Through Violence,* 272.
39. Ibid, 77.
40. Alexander Saxton, *The Rise and Fall of the White Republic,* 72–77.
41. Slotkin, *Regeneration Through Violence,* 398.
42. Smith, *Virgin Land,* 120.
43. Sale, *The Conquest of Paradise,* 19.
44. Phillips and Phillips, *The Worlds of Christopher Columbus,* 9.
45. Sale, *The Conquest of Paradise,* 222.
46. Phillips and Phillips, *The Worlds of Christopher Columbus,* 5.
47. Sale, *The Conquest of Paradise,* 340.

48. Ibid, 360.
49. Marcus Cunliffe, 'Introduction' to *The Life of Washington.*
50. Boorstin, *The National Experience,* 359–361.
51. Ibid, 382.
52. 'The Significance of the Frontier in American History', 46.
53. Dale Carter, *The Final Frontier,* 94.
54. *The Image,* 54–55.
55. Carter, *The Final Frontier,* 157, 188.
56. 'Head of NASA Has a New Vision of 1984', *The New York Times,* 17 July 1969, 47.
57. Edmundo O'Gorman, *The Invention of America.*
58. Ibid.
59. Ibid, 129.
60. Walter LaFeber, *The New Empire.*
61. Alan Henrikson, 'Maps, Globes, and the "Cold War"', 447.
62. Drinnon, *Facing West,* 369.
63. See Perry Miller, *Errand into the Wilderness,* John Hellman, *American Myth and the Legacy of Vietnam,* Philip Melling, *Vietnam in American Literature.*
64. John Hellman, *American Myth and the Legacy of Vietnam,* 36.
65. *A Rumour of War,* xii.
66. Ibid, 4.
67. Ibid, 5.
68. *Why are we in Vietnam?,* 122.
69. *Deliverance,* 3.
70. Hellman, *American Myth and the Legacy of Vietnam,* 45.
71. James William Gibson, *The Perfect War,* 33.
72. Benedict Anderson, *Imagined Communities,* 158.
73. Frances Fitzgerald, *Fire in the Lake,* 52.
74. Donald Ringnalda, 'Fighting and Writing: America's Vietnam War Literature'.
75. Gibson, *The Perfect War,* 85.
76. *A Rumour of War,* 95.
77. Tim O'Brien, *The Things They Carried,* 31–32.
78. Stephen Wright, *Meditations in Green,* 146.
79. 'The cultural logic of late capitalism', 44.
80. *A Rumour of War,* 96.
81. *American Myth and the Legacy of Vietnam,* 209.
82. Ibid, 217.
83. Dilip Hiro, *Desert Shield to Desert Storm: The Second Gulf War,* 374–378.
84. Gibson, *The Perfect War,* 456.
85. 'The Last Crusade', 114.
86. *Dispatches,* 144.
87. H. Bruce Franklin, *M.I.A. or Mythmaking in America,* 133–134.
88. 'The Significance of the Frontier in American History', 61.
89. 'The Problem of the West', 63.
90. Drinnon, *Facing West,* 465.
91. 'Can a dead world be brought to life? The greening of the red planet may be the next giant step for mankind', *Life,* May 1991, 32.
92. Ibid, 34.

93. The biosphere project eventually ran into difficulties and acrimony among the participants, along with an admission by the organizers that oxygen had been pumped in after the dome had been sealed.

CHAPTER 7: THE IMPERIALIST MAP: BEYOND MATERIALISM AND IDEALISM

1. *Heart of Darkness*, 33.
2. Quoted in T.C. McLuhan, *Touch the Earth*, 54.
3. Quoted by Walter McDougall, *The Heavens and the Earth*, 239.
4. See, for example, Antonio Gramsci, *Prison Notebooks*; Louis Althusser, *For Marx*; Nicos Poulantzas, *State, Power, Socialism*.
5. *The Question of Palestine*, 73–74.
6. Edward Said, *Culture and Imperialism*.
7. *The Question of Palestine*, 86.
8. Joshua, xvii, 14–18.
9. *Discourse on Method*, 56.
10. Foucault, *The Order of Things*, 55–56.
11. Michael Adas, *Machines as the Measure of Men*, 24–32; J.H. Parry, *The Age of Reconnaissance*, 37, 100; Francis Jennings, *The Invasion of America*, 3.
12. Adas, *Machines as the Measure of Men*, 7.
13. Jennings, *The Invasion of America*.
14. For illustrations, see Robert Lister and Florence Lister, *Those Who Came Before*, 21; Edward Dozier, *The Pueblo Indians of North America*, 33.
15. Rik Pinxten, *Anthropology of Space*; Ray Williamson, *Living in the Sky*.
16. Leslie White, *The Pueblo of Santa Ana, New Mexico*; Dozier, *The Pueblo Indians of North America*; Williamson, *Living in the Sky*.
17. George Bird Grinnell, *The Cheyenne Indians: Their History and Ways of Life*, 88, 94; Royal B. Hassrick, *The Sioux: Life and Customs of a Warrior Society*, 206; John Neihardt (ed.), *Black Elk Speaks*, 2.
18. Grinnell, *The Cheyenne Indians*; E. Adamson Hoebel, *The Cheyennes*; Hassrick, *The Sioux*; James Walker, *Lakota Belief and Ritual*; Neihardt (ed.), *Black Elk Speaks*; Williamson, *Living in the Sky*.
19. Benjamin Lee Whorf, 'An American Indian Model of the Universe'.
20. See Kirkpatrick Sale, *The Conquest of Paradise*, 78–80.
21. Peter Matthiessen, *In the Spirit of Crazy Horse*.
22. Lewis Mumford, *Technics and Civilization*, 49.
23. Frederick Turner, *Beyond Geography*, 92.
24. Robert Berkhofer, *The White Man's Indian*, 120–125.
25. *National Geographic*, October 1991.
26. Roderick Nash, *Wilderness and the American Mind*, 44.
27. There are, of course, numerous traces left of the earliest indigenous peoples, spectacular examples including the ancient Anasazi ruins in Canyon de Chelly, Arizona.
28. Berkhofer, *The White Man's Indian*, 86.
29. *The Americans: The Colonial Experience*, 149, 150.
30. *American Foreign Policy*, 48.

31. *The Perfect War*, 16, original emphasis.
32. *American Foreign Policy*, 57.
33. Ibid, 54.
34. Andrew Wilson, *The Bomb and the Computer*, 2.
35. Ibid, 20.
36. Ibid.
37. Howard Rheingold, *Virtual Reality*, 45.
38. Paul Edwards, 'The Closed World', 153.
39. Gabriel Kolko, *Vietnam: Anatomy of a War*, 60.
40. *The Perfect War*, 18.
41. Ibid, 80–81; Walt Rostow, *The Stages of Growth*.
42. Loren Baritz, *Backfire*, 257–258.
43. Gibson, *The Perfect War*, 123.
44. See for example, Neil Sheehan, *A Bright Shining Lie*, 712; Frances Fitzgerald, *Fire in the Lake*, 535.
45. Gibson, *The Perfect War*, 396.
46. Ibid, 292; see also Fitzgerald, *Fire in the Lake*, 454–456.
47. Baritz, *Backfire*, 164.
48. Sheehan, *A Bright Shining Lie*, 323–324.
49. Gibson, *The Perfect War*, 308.
50. Ibid, 435.
51. See, for example, Philip Beidler, *American Literature and the Experience of Vietnam*, Thomas Myers, *Walking Point: American Narratives of Vietnam*.
52. Andrew Rotter, *The Path to Vietnam*; William Borden, *The Pacific Alliance*.
53. Chomsky, 'Vietnam and United States Global Strategy', 240.
54. Chomsky, 'Punishing Vietnam', 301.
55. Chomsky, 'Intervention in Vietnam and South America: Parallels and Differences', 325; Noam Chomsky and Edward Herman, *After the Cataclysm*.
56. William Shawcross, *Sideshow*, 271–272.
57. Stephen Burman, *America in the Modern World*, 36.
58. Kirkpatrick Sale, *Power Shift*.
59. *Marvellous Possessions*, 64.
60. Henri Lefebvre, *The Production of Space*, 151.
61. Ibid, 152.
62. 'Popular-Democratic vs Authoritarian Populism: Two Ways of "Taking Democracy Seriously"', 143.
63. *Maps and Dreams*.
64. V.I. Lenin, *Materialism and Empirio-Criticism*, 20.

CHAPTER 8: DECONSTRUCTING THE MAP

1. 'Signature, Event, Context', 329.
2. Ibid, 330.
3. Derrida, 'Outwork', 6.
4. Derrida, *Spurs: Nietzsche's Styles*, 51.
5. *Gynesis*, 25, 52.

6. 'Sorties: Out and Out: Attacks/Ways Out/Forays', 68.
7. *This Sex Which Is Not One*, 213; see also 106–114, and *Speculum of the Other Woman.*
8. Robert Bly, *Iron John.*
9. *The Lay of the Land*, 88.
10. 'Women in the Beehive: A Seminar with Jacques Derrida', 94.
11. Suzanne Moore, 'Getting a Bit of the Other – the Pimps of Postmodernism'.
12. *The Robber Bride*, 464.
13. *Gödel, Escher, Bach.*
14. William Boelhower, *Through a Glass Darkly*, 71.
15. Graham Huggan, 'Decolonizing the Map', 117.
16. Ibid, 119.
17. José Rabasa, 'Allegories of the *Atlas*', 11.
18. Huggan, 'Decolonizing the Map', 121.
19. *Ragtime*, 19.
20. Mark Monmonier, *How to Lie with Maps*, 141–142.
21. John Shepherd, 'Poverty and Comfort', in Barber and Board *Tales from the Map Room*, 146–147.
22. Dilip Hiro, *Desert Shield to Desert Storm*, 120–121.
23. Arthur Cracknell and Ladson Hayes, *Introduction to Remote Sensing*, 171, 186–188.
24. Simon Berthon and Andrew Robinson, *The Shape of the World*, 171–185; Cracknell and Hayes, *Introduction to Remote Sensing.*
25. Peter Matthiessen, *In the Spirit of Crazy Horse.*
26. Berthon and Robinson, *The Shape of the World*, 177.
27. Mark Collins, 'Mapping the tropical forests'.
28. Kevin Robins, 'Into the Image', 72.
29. *The Non-Objective World*, 68.
30. Gertrude Stein, *Picasso*, 50.
31. *The Fractal Geometry of Nature*, 26.
32. Lewis Richardson, 1961; cited by Mandelbrot in *The Fractal Geometry of Nature*, 27.
33. *The Fractal Geometry of Nature*, 5.
34. 'Secondary Elaborations', 416.

Bibliography

Abish, Walter, *Alphabetical Africa* (New York, 1974).
Adams, Percy, *Travellers and Travel Liars, 1660–1800*, 1962 (New York, 1980).
Adas, Michael, *Machines as the Measure of Men: Science, Technology and Ideologies of Western Dominance* (Ithaca, NY, 1989).
Ager, John, 'Maps and Propaganda', *Bulletin of the Society of University Cartographers*, vol. 11 (1977).
Allen, John, 'Lands of Myth, Waters of Wonder: The Place of the Imagination in the History of Geographical Exploration', in *Geographies of the Mind: Essays in Historical Geosophy*, eds David Lowenthal and Martyn Bowden (Oxford, 1976).
Althusser, Louis, *For Marx*, 1965, trans. Ben Brewster (London, 1977).
Ambrose, Stephen, *Rise to Globalism: American Foreign Policy Since 1938* (Harmondsworth, 1991).
Anderson, Benedict, *Imagined Communities: Reflections on the Origin and Spread of Nationalism*, revised edition (London, 1991).
Ascherson, Neal, 'The new Europe', *The Independent on Sunday*, Review, 9 February 1992.
Atwood, Margaret, *The Robber Bride* (London, 1993).
Bagrow, Leo, *History of Cartography*, 1944, trans. D.L. Paisey, 1960; revised and enlarged by R.A. Skelton (Chicago, 1985).
Barber, Peter and Board, Christopher, *Tales from the Map Room: Fact and Fiction about Maps and Their Makers* (London, 1993).
Baritz, Loren, *Backfire: A history of how American culture led us into Vietnam and made us fight the way we did* (New York, 1985).
Barthes, Roland, *Mythologies*, 1957, trans. Annette Lavers (London, 1973).
Baudrillard, Jean, 'Consumer Society', 1970, in *Selected Writings*, ed. Mark Poster (Oxford, 1988).
—— *For a Critique of the Political Economy of the Sign*, 1972, trans. Charles Levin (St Louis, Miss., 1981).
—— *Simulations*, 1981, trans. Paul Foss, Paul Patton and Philip Beitchman (New York, 1983).
—— 'The Ecstasy of Communication', 1983, trans. John Johnston, in *Postmodern Culture*, ed. Hal Foster (London, 1985).
—— *America*, 1986, trans. Chris Turner (London, 1988).
—— 'Panic Crash', trans. Faye Trecartin and Arthur Kroker, in *Panic Encyclopedia*, eds Arthur Kroker, Marilouise Kroker and David Cook (London, 1989).
—— *La guerre du Golfe n'a pas eu lieu* (Paris, 1991).
—— 'The Reality Gulf', in *The Guardian*, 1 January 1991.
Beazley, C. Raymond, *The Dawn of Modern Geography* (London, 1897).
Beer, Gillian, *Darwin's Plots* (London, 1983).
Beidler, Philip, *American Literature and the Experience of Vietnam* (Athens, Ga., 1982).

Benjamin, Walter, 'The Work of Art in the Age of Mechanical Reproduction', 1936, in *Illuminations*, ed. Hannah Arendt (London, 1970).
Berkhofer, Robert, *The White Man's Indian: Images of the American Indian from Columbus to the Present* (New York, 1978).
Berndt, Ronald and Berndt, Catherine, *The Speaking Land: Myth and Story in Aboriginal Australia* (Victoria, Australia, 1988).
Berthon, Simon and Robinson, Andrew, *The Shape of the World* (London, 1991).
Biskind, Peter, 'The Last Crusade', in *Seeing Through Movies*, ed. Mark Crispin Miller (New York, 1990).
Blakemore, M.J. and Harley, J.R., 'Concepts in the History of Cartography: A Review and Perspective', *Cartographica*, vol. 17, December 1980.
—— 'Concepts in the History of Cartography: Questions of Definiton', *Cartographica*, vol. 19, December 1982.
Bly, Robert, *Iron John: A Book About Men* (Shaftesbury, Dorset, 1993).
Bode, Steven and Wombell, Paul, 'In a new light', introduction to *Photovideo: Photography in the age of the computer* (London, 1991).
Boelhower, William, *Through A Glass Darkly: Ethnic Semiosis in American Literature* (New York, 1987).
Boggs, S.W., 'Cartohypnosis', *Scientific Monthly*, vol. 64, October 1947.
Boorstin, Daniel, *The Americans: The Colonial Experience*, 1958 (London, 1988).
—— *The Image: A Guide to Pseudo-Events in America*, originally published as *The Image, or What Happened to the American Dream*, 1961 (New York, 1987).
—— *The Americans: The National Experience*, 1965 (London, 1988).
—— *The Americans: The Democratic Experience*, 1973 (London, 1988).
Borden, William, *The Pacific Alliance: United States Foreign Economic Policy and Japanese Trade Recovery* (Madison, Wis., 1984).
Borges, Jorge Luis, 'Of Exactitude in Science', in *A Universal History of Infamy*, 1954, trans. Norman Thomas di Giovanni (Harmondsworth, 1975).
Bourdieu, Pierre, *Outline of a Theory of Practice*, 1972, trans. Richard Nice (Cambridge, 1977).
Bradford, William, *Of Plymouth Plantation*, extracted in *The American Puritans: Their Prose and Poetry*, ed. Perry Miller (New York, 1956).
Brecht, Bertolt, 'On the Formalistic Character of the Theory of Realism', trans. Stuart Hood, in *Aesthetics and Politics* (London, 1980).
Brody, Hugh, *Maps and Dreams* (London, 1986).
Burman, Stephen, *America in the Modern World: The Transcendence of United States Hegemony* (Hemel Hempstead, Herts., 1991).
Calvino, Ital, *Invisible Cities*, 1972, trans. William Weaver (London, 1979).
Canary, Robert and Kozicki, Henry, *The Writing of History* (Madison, Wis., 1978).
Caputo, Philip, *A Rumour of War* (London, 1978).
Carey, Peter, 'Do You Love Me?', in *Exotic Pleasures* (London, 1981).
Carroll, Lewis, *Sylvie and Bruno Concluded* (London, 1894).
Carter, Dale, *The Final Frontier: The Rise and Fall of the American Rocket State* (London, 1988).
Carter, Paul, *The Road to Botany Bay* (London, 1987).
Chaliand, Gérard and Rageau, Jean-Pierre, *Strategic Atlas: World Geopolitics* (Harmondsworth, 1983).

Chiapelli, Fred, ed., *First Images of America: The Impact of the New World on the Old* (Berkeley, Calif., 1976).
Chomsky, Noam, 'Vietnam and United States Global Strategy', 1973, in *The Chomsky Reader*, ed. James Peck (London, 1988).
—— 'Punishing Vietnam', 1982, in *The Chomsky Reader*.
—— 'Intervention in Vietnam and Central America: Parallels and Differences', 1985, in *The Chomsky Reader*.
—— *U.S. Gulf Policy*, Open Magazine Pamphlet Series, March 1991.
—— *The New World Order*, Open Magazine Pamphlet Series, August 1991.
—— *Deterring Democracy* (London, 1992).
Chomsky, Noam and Herman, Edward, *After the Cataclysm: Postwar Indochina and the Reconstruction of Imperial Ideology* (Boston, 1979).
Cixous, Hélène, 'Sorties: Out and Out: Attacks/Ways Out/Forays', in Hélène Cixous and Catherine Clément, *The Newly Born Woman*, 1985, trans. Betsy Wing (Manchester, 1986).
Collins, Mark, 'Mapping the tropical forests', *Mapping Awareness*, vol. 5, 10 December 1991.
Columbus, Christopher, *The Journal of Christopher Columbus*, trans. Cecil Jane, revised L.A. Vigneras (London, 1960).
—— 'Narrative of the Third Voyage of Columbus to the Indies, in which he discovered the mainland, dispatched to the Sovereigns from the island of Hispaniola', ed. and trans. J.M. Cohen, in *The Four Voyages of Christopher Columbus* (Harmondsworth, 1969).
Columbus, Hernando, 'Life of the Admiral by his son, Hernando Colon', extracts in *The Four Voyages of Christopher Columbus*, ed. and trans. J.M. Cohen, (Harmondsworth, 1969).
Conrad, Joseph, *Heart of Darkness*, 1902 (Harmondsworth, 1983).
Cooper, David, *Metaphor* (London, 1986).
Cortés, Hernán, *Letters from Mexico*, ed. and trans. Anthony Pagden (New Haven, Conn., 1986).
Cosgrove, Denis, 'The myth and the stones of Venice: an historical geography of a symbolic landscape', *Journal of Historical Geography*, vol. 8, no. 2, 1982.
—— 'Prospect, Perspective and the Evolution of the Landscape Idea', *Transactions*, Institute of British Geographers (New Series), vol. 10, December 1987.
Cracknell, Arthur and Hayes, Ladson, *Introduction to Remote Sensing* (London, 1991).
Crèvecoeur, J. Hector St John, *Letters from an American Farmer*, 1801, ed. and trans. Percy Adams (Harmondsworth, 1986).
Crockett, David, *A Narrative of the Life of David Crockett*, 1834, ed. Joseph Arpad (New Haven, Conn., 1972).
Crone, G.R., *Maps and their Makers* (London, 1962).
Cronon, William, *Nature's Metropolis: Chicago and the Great West* (New York, 1991).
Cumming, W.P., Skelton, R.A. and Quinn, D.B., *The Discovery of North America* (London, 1971).
Cunliffe, Marcus, 'Introduction to Mason Weems', *The Life of Washington* (Cambridge, Mass., 1962).
Darrach, Brad and Petranek, Steve, 'Can a dead world be brought to life?

The greening of the red planet may be the next giant step for mankind', *Life*, May 1991.
Davis, Mike, *City of Quartz: Excavating the Future in Los Angeles* (London, 1992).
Debord, Guy, *Society of the Spectacle*, 1967, trans. (London, 1987).
—— *Comments on the Society of the Spectacle*, 1988, trans. Malcolm Imrie (London, 1990).
Deleuze, Gilles and Guattari, Felix, *Anti-Oedipus: Capitalism and Schizophrenia*, 1972, trans. Robert Hurley, Mark Seem and Helen Lane (London, 1984).
—— *On the Line*, trans. John Johnston (New York, 1983).
Del Vecchio, John, *The 13th Valley* (London, 1983).
Derrida, Jacques, *Speech and Phenomena: And Other Essays on Husserl's Theory of Signs*, 1967, trans. David Allison (Evanston, Wyo., 1973).
—— 'Signature, Event, Context', in *Margins of Philosophy*, 1972, trans. Alan Bass (Brighton, E. Sussex, 1982).
—— 'White Mythology: Metaphor in the Text of Philosopy', in *Margins of Philosophy*.
—— 'Outwork, prefacing', in *Dissemination*, 1972, trans. Barbara Johnson (Chicago, 1981).
—— *Spurs: Nietzsche's Styles*, 1978, trans. Barbara Harlow (Chicago, 1978).
—— 'Living on/Borderlines', in *Deconstruction and Criticism*, eds H. Bloom *et al.* (London, 1979).
—— 'The Law of Genre', trans. Avital Ronnell, *Critical Inquiry*, vol. 7, Autumn 1980.
—— 'Women in the Beehive: A Seminar with Jacques Derrida', 1984, in *Men in Feminism*, eds Alice Jardine and Paul Smith (New York, 1987).
Descartes, René, *Discourse on the Method of Properly Conducting One's Reason and of Seeking the Truth in the Sciences*, 1637, trans. E. Sutcliffe (Harmondsworth, 1968).
—— *Meditations*, 1641, trans. E. Sutcliffe (Harmondsworth, 1968).
De Vorsey, Louis, 'Amerindian contributions to the mapping of America: a preliminary view', *Imago Mundi*, vol. 30, 1978.
Díaz, Bernal, *The Conquest of New Spain*, trans. J.M. Cohen, (Harmondsworth, 1963).
Dickens, Charles, *Martin Chuzzlewit*, 1843–4 (Harmondsworth, 1986).
Dickey, James, *Deliverance* (London, 1970).
Diski, Jenny, *Nothing Natural* (London, 1987).
—— *Rainforest* (Harmondsworth, 1988).
Doctorow, E.L., *Ragtime* (New York, 1974).
—— 'False Documents', in *E.L. Doctorow, Essays and Conversations*, ed. R. Trenner (Princeton, NJ, 1983).
Douglas, Mary, *Purity and Danger*, 1966 (London, 1984).
—— *Natural Symbols* (London, 1973).
Dozier, Edward, *The Pueblo Indians of North America* (New York, 1970).
Drinnon, Richard, *Facing West: The Metaphysics of Indian Hating and Empire Building* (New York, 1980).
Eagleton, Terry, 'Capitalism, Modernism and Postmodernism', in *Against the Grain, Selected Essays* (London, 1986).
Eco, Umberto, 'Travels in Hyperreality', 1973, trans. William Weaver (London, 1987).

Edgerton, Samuel, *The Renaissance Rediscovery of Linear Perspective* (New York, 1975).
Edwards, Paul, Levidow, Les and Robins, Kevin, eds, 'The Closed World: Systems discourse, military policy and post-World War II US historical consciousness, in *Cyborg Worlds: The Military Information Society* (London, 1989).
Evans-Pritchard, E., *Witchcraft, Oracles and Magic among the Azande* (Oxford, 1976).
Featherstone, Mike, *Consumer Culture and Postmodernism* (London, 1991).
Filson, John, *The Discovery, Settlement and present state of Kentucke*, 1784 (New York, 1962).
Fitzgerald, Frances, *Fire in the Lake: The Vietnamese and the Americans in Vietnam* (New York, 1971).
Foucault, Michel, *The Order of Things: An Archaeology of the Human Sciences*, 1966, trans. (London, 1970).
—— *Discipline and Punish: The Birth of the Prison*, 1975, trans. Alan Sheridan (London, 1979).
Franklin, H. Bruce, *M.I.A. or Mythmaking in America* (New York, 1992).
Friel, Brian, *Translations*, 1980, in *Selected Plays of Brian Friel* (London, 1984).
Geertz, Clifford, 'Common Sense as Cultural System', in *Local Knowledge: Further Essays in Interpretive Anthroplogy* (New York, 1983).
Gibson, James, *The Perfect War: Technowar in Vietnam* (Boston, 1986).
Gleick, James, *Chaos: Making a New Science* (London, 1988).
Glenny, Misha, *The Fall of Yugoslavia: The Third Balkan War* (Harmondsworth, 1993).
Goffman, Erving, *Frame Analysis: An Essay on the Organization of Experience*, 1974 (Boston, 1986).
Gombrich, Ernst, *The Sense of Order* (Oxford, 1979).
Gould, Peter and White, Rodney, *Mental Maps* (Boston, 1986).
Gramsci, Antonio, *Selections from the Prison Notebooks*, ed. and trans. Quintin Hoare and Geoffrey Nowell Smith (London, 1971).
Greenblatt, Stephen, *Marvellous Possessions: The Wonder of the New World* (Oxford, 1991).
Grinnell, George Bird, *The Cheyenne Indians: Their History and Ways of Life* (New York, 1962).
Hale, J.R., *Renaissance Europe* (London, 1971).
Hall, Stuart, 'Popular-Democratic vs Authoritarian Populism: Two Ways of "Taking Democracy Seriously"', 1980, in *The Hard Road to Renewal: Thatcherism and the Crisis of the Left* (London, 1988).
Hamer, Mary, 'Putting Ireland on the Map', *Textual Practice*, vol. 3, no. 3, Summer 1989.
Harley, J.B., 'Meaning and ambiguity in Tudor cartography', in *English Map-Making 1500–1650, Historical Essays* (London, 1983).
Hart, Kevin, 'Maps of Deconstruction', *Meanjin*, vol. 45, no. 1, 1986.
Harvey, David, *The Condition of Postmodernity* (Oxford, 1989).
Harvey, P.D.A., *The History of Topographical Maps, Symbols, Pictures and Surveys* (London, 1980).
Hassrick, Royal, *The Sioux: Life and Customs of a Warrior Society* (Norman, Okla., 1964).

Hawthorne, Nathaniel, *The Scarlet Letter*, 1850 (Harmondsworth, 1970).
Hebdige, Dick, *Hiding in the Light* (London, 1988).
Helgerson, R., 'The Land Speaks: Cartography, Chorography, and Subversion in Renaissance England', *Representations*, vol. 16, Fall 1986.
Hellman, John, *American Myth and the Legacy of Vietnam* (New York, 1986).
Henrikson, Alan, 'Maps, Globes, and the "Cold War"', in *Special Libraries*, October/November 1974.
Herr, Michael, *Dispatches* (London, 1978).
Herring, George, *America's Longest War: The United States and Vietnam 1950–1975* (New York, 1986).
Hiro, Dilip, *Desert Shield to Desert Storm: The Second Gulf War* (London, 1992).
Hobsbawm, Eric and Ranger, Terrence, eds, *The Invention of Tradition* (Cambridge, 1984).
Hodgkiss, A.G., *Understanding Maps* (Folkestone, Kent, 1981).
Hoebel, E. Adamson, *The Cheyennes: Indians of the Great Plains* (New York, 1960).
Hofstadter, Douglas, *Gödel, Escher, Bach: An Eternal Golden Braid* (Harmondsworth, 1980).
Huggan, Graham, 'Decolonizing the Map', *Ariel*, vol. 20, no. 4, 1989.
Hutcheon, Linda, *A Poetics of Postmodernism: History, Theory, Fiction* (London, 1988).
—— *The Politics of Postmodernism* (London, 1989).
International Affairs (Moscow), 'Leading on Earth and in Space' (editorial), vol. 8, no. 9, 1962.
Irigaray, Luce, *Speculum of the Other Woman*, 1974, trans. Gillian Gill (Ithaca, NY, 1985).
—— *This Sex Which Is Not One*, 1977, trans. Catherine Porter, (Ithaca, NY, 1985).
Jackson, J.B., 'The Order of a Landscape: Reason and Religion in Newtonian America', in *The Interpretation of Ordinary Landscapes*, ed. D.W. Meinig (New York, 1979).
Jackson, Peter, *Maps of Meaning: An Introduction to Cultural Geography* (London, 1989).
James, Preston and Martin, Geoffrey, *All Possible Worlds: A History of Geographical Ideas* (New York, 1972).
Jameson, Fredric, *The Prison-House of Language* (Princeton, NJ, 1972).
—— 'Cognitive Mapping', 1983, in *Marxism and the Interpretation of Culture*, eds Cary Nelson and Lawrence Grossberg (London, 1988).
—— 'The Cultural Logic of Late Capitalism', 1984, in *Postmodernism, or, The Cultural Logic of Late Capitalism* (London, 1991).
—— 'Secondary Elaborations', in *Postmodernism, or, The Cultural Logic of Late Capitalism* (London, 1991).
Jardine, Alice, *Gynesis: Configurations of Woman and Modernity* (Ithaca, NY, 1985).
Jefferson, Thomas, *Notes on the State of Virginia*, 1861 (New York, 1964).
Jenks, Charles, 'Stone, Paper, Scissors', *Marxism Today*, February 1991.
Jennings, Francis, *The Invasion of America: Indians, Colonialism, and the Cant of Conquest* (New York, 1976).

Jervis, W.W., *The World in Maps: A Study in Map Evolution* (London, 1936).
Johnson, Hildegard, *Order Upon the Land: The U.S. Rectangular Survey and the Upper Mississippi Country* (Oxford, 1976).
Jolliffe, R., 'An Information Theory Approach to Cartography', *Cartography*, vol. 8, October 1974.
Jones, Howard Mumford, *O Strange New World: American Culture, the Formative Years*, 1952 (New York, 1964).
Kaplan, E. Ann, *Rocking Around the Clock: Music Television, Postmodernism, and Consumer Culture* (New York, 1987).
Keillor, Garrison, *Lake Wobegone Days* (London, 1987).
Kennedy, John F., 'Special Message to the Congress on Urgent National Needs', in *Public Papers of the Presidents of the United States: John F. Kennedy* (Washington, DC, 1962).
Kern, Stephen, *The Culture of Time and Space: 1880–1918* (London, 1983).
Kidron, Michael and Segal, Ronald, *The New State of the World Atlas* (London, 1991).
Kissinger, Henry, *American Foreign Policy*, 1969 (New York, 1977).
Kolko, Gabriel, *Vietnam: Anatomy of a War, 1940–1975* (London, 1986).
Kolodny, Annette, *The Lay of the Land: Metaphor as Experience and History in American Life and Letters* (Chapel Hill, NC, 1975).
Krauss, Rosalind, 'The Originality of the Avant-Garde', in *The Originality of the Avant-Garde and other Modernist Myths* (Cambridge, Mass., 1981)
Kroker, Arthur and Cook, David, *The Postmodern Scene: Excremental Culture and Hyper-Aesthetics* (London, 1988).
Kroker, Arthur, Kroker, Marilouise and Cook, David, *Panic Encyclopedia* (London, 1989).
La Feber, Walter, *The New Empire: An Interpretation of American Expansion 1860–1898* (Ithaca, NY, 1963).
Lakoff, George and Johnson, Mark, *Metaphors We Live By* (Chicago, 1980).
Lash, Scott, *Sociology of Postmodernism* (London, 1990).
Latham, Ronald, 'Introduction to Marco Polo', *The Travels of Marco Polo* (Harmondsworth, 1958).
Laurel, Brenda, ed., *The Art of Human-Computer Interface Design* (Reading, Mass., 1990).
Leach, Edmund, *Culture and Communication* (Cambridge, 1976).
Lefebvre, Henri, *The Production of Space*, 1974, trans. Donald Nicholson-Smith (Oxford, 1991).
Lenin, V.I., *Materialism and Empirio-Criticism: Critical Comments on Reactionary Philosophy*, 1908, trans. (Moscow, 1970).
Levidow, Les and Robins, Kevin, *Cyborg Worlds: The Military Information Society* (London, 1989).
Lévi-Strauss, Claude, *The Elementary Structures of Kinship*, 1949/1967, trans. J. Bell, J. von Sturmer and R. Needham (London, 1969).
—— *Totemism*, 1962, trans. R. Needham (Boston, 1963).
—— *The Savage Mind*, 1962, trans. (London, 1972).
Lifton, Robert, *Boundaries* (New York, 1969).
Lister, Robert and Lister, Florence, *Those Who Came Before: Southwestern Archaeology in the National Park System* (Tucson, Ariz., 1983).

Logsdon, John, *The Decision to Go to the Moon: Project Apollo and the National Interest* (Boston, 1970).
Lynch, Kevin, *The Image of the City* (Cambridge, Mass., 1960).
Lyotard, Jean-François, *The Postmodern Condition: A Report on Knowledge*, 1979, trans. Geoff Bennington (Manchester, 1986).
MacCannell, Dean, *The Tourist: A New Theory of the Leisure Class* (London, 1976).
McCormick, Thomas, *America's Half-Century: United States Foreign Policy in the Cold War* (Baltimore, Va., 1989).
McDougall, Walter, *The Heavens and the Earth: A Political History of the Space Age* (New York, 1984).
McEwan, Ian, *The Comfort of Strangers* (London, 1982).
McKee, Victoria, 'The ultimate stay at home holiday – with computer technology, travellers can avoid the nasty bits of "abroad"', *The Guardian*, 16 February 1991.
Mackinder, Halford, 'The Geographical Pivot of History', *The Geographical Journal*, vol. 23, 4 April 1904.
McLuhan, Marshall, *Understanding Media*, 1964 (London, 1967).
McLuhan, T.C., ed., *Touch the Earth: A Self-Portrait of Indian Existence* (London, 1973).
Magas, Branka, *The Destruction of Yugoslavia: Tracking the Break-Up 1980–1992* (London, 1993).
Mailer, Norman, *Why are we in Vietnam?*, 1967 (Oxford, 1988).
—— *A Fire on the Moon* (London, 1970).
Malevich, Kasimir, *The Non-Objective World*, 1927, trans. Howard Dearstyne (Chicago, 1959).
Mandel, Ernest, *Late Capitalism*, 1972, trans. Joris De Bres (London, 1975).
Mandelbrot, Benoit, *The Fractal Geometry of Nature*, revised edition (New York, 1983).
Mandeville, John, *The Travels of Sir John Mandeville*, trans. C.W.R.D. Moseley (Harmondsworth, 1983).
Martin, Rupert, ed., *The View from Above: 125 Years of Aerial Photography* (London, 1983).
Marx, Karl, *The Communist Manifesto*, 1872, trans. Samuel Moore (Harmondsworth, 1967).
Marx, Leo, *The Machine in the Garden; Technology and the Pastoral Ideal in America* (Oxford, 1964).
Matthiessen, Peter, *In the Spirit of Crazy Horse*, 1983 (London, 1991).
Mead, Robin, 'Treason on a big scale', *Weekend Guardian*, 3–4 November 1990.
Melling, Philip, *Vietnam in American Literature* (Boston, 1990).
Miller, Perry, *The American Puritans: Their Prose and Poetry* (New York, 1956).
—— *Errand into the Wilderness* (Cambridge, Mass., 1956).
Monmonier, Mark, *Maps, Distortion, and Meaning*, Resource Paper No. 75, Association of American Geographers (New York, 1977).
—— *How to Lie with Maps* (Chicago, 1991).
Moore, Suzanne, 'Getting a Bit of the Other – the Pimps of Postmodernism', in *Male Order: Unwrapping Masculinity*, eds Rowena Chapman and Jonathan Rutherford (London, 1988).

More, Thomas, *Utopia*, 1516, trans. Paul Turner (Harmondsworth, 1965).
Morgan, Victor, 'The Cartographic Image of "The Country" in Early Modern England', in *Transactions of the Royal Historical Society*, 5th series, vol. 29 (London, 1979).
Muehrcke, Philip, 'Map Reading and Abuse', *Journal of Geography*, May 1974.
Mumford, Lewis, *Technics and Civilization* (London, 1934).
Muntz, Philip, 'Union Mapping in the American Civil War', *Imago Mundi*, vol. XVII, 1963.
Myers, Thomas, *Walking Point: American Narratives of Vietnam* (New York and Oxford, 1988).
Nairn, Tom, *The Enchanted Glass: Britain and its Monarchy* (London, 1988).
Nash, Roderick, *Wilderness and the American Mind*, 1967 (New Haven, Conn., 1982).
Neihardt, John, ed., *Black Elk Speaks* (Lincoln, Nebr., 1979).
New York Times, 'Barbarism with Sputniks' (editorial), 4 November 1957.
—— 'Soviet Distorted Recent Maps; Security Believed the Reason', 18 January 1970.
Nietzsche, Friedrich, 'On Truth and Falsity in their Ultramoral Sense', 1873, trans. M. Mugge, in *Early Greek Philosopy and Other Essays*, vol. 2 of the *Complete Works of Friedrich Nietzsche*, ed. Oscar Levy (London, 1911).
O'Brien, Tim, *The Things They Carried* (London, 1991).
O'Donaghue, Yolande, *William Roy (1726–1790): Pioneer of the Ordnance Survey* (London, 1977).
O'Gorman, Edmundo, *The Invention of America*, 1961 (Westport, Conn., 1972).
Orvell, Miles, *The Real Thing: Imitation and Authenticity in American Culture, 1880–1940* (Chapel Hill, NC, 1989).
Packard, Vance, *The Hidden Persuaders*, 1957 (Harmondsworth, 1981).
Paine, Thomas, 'Head of NASA Has a New Vision of 1984', *The New York Times*, 17 July 1969.
Panofsky, Erwin, *Meaning in the Visual Arts* (New York, 1955).
Parry, J.H., *The Age of Reconnaissance* (London, 1963).
Peters, Arno, 'Foreword', *Peters Altas of the World* (Harlow, Essex, 1989).
Phillips, William and Phillips, Carla, *The Worlds of Christopher Columbus* (Cambridge, 1992).
Pinxten, Rik, *Anthropology of Space: Explorations in the Natural Philosophy and Semantics of the Navajo* (Philadelphia, 1983).
Pliny the Elder, *Natural History: A Selection*, trans. John Healy (Harmondsworth, 1991).
Pohl, Frederick, *Amerigo Vespucci, Pilot Major* (New York, 1966).
Polo, Marco, *The Travels of Marco Polo*, trans. Ronald Latham (Harmondsworth, 1958).
Poulantzas, Nicos, *State, Power, Socialism*, 1978, trans. Patrick Camiller (London, 1978).
Prigogine, Ilya and Stengers, Isabelle, *Order out of Chaos: Man's new dialogue with nature* (London, 1985).
Rabasa, José, 'Allegories of the *Atlas*', in eds, Frances Barker, *et al., Europe and its Others* (Exeter, Devon, 1984).
Raisz, Erwin, *General Cartography* (New York, 1948).

Rees, Ronald, 'Historical Links Between Cartography and Art', *Geographical Review*, vol. 70, 1980.
Reps, John, *Town Planning in Frontier America* (Princeton, NJ, 1969).
—— *Cities of the American West: A History of Frontier Urban Planning* (Princeton, NJ, 1979).
Rheingold, Howard, *Virtual Reality* (London, 1991).
Ricoeur, Paul, *The Rule of Metaphor*, 1975, trans. Robert Czerny (London, 1986).
Ringnalda, Donald, '"Fighting and Writing": America's Vietnam War Literature', in *Journal of American Studies*, vol. 22, no. 1, 1988.
Ritchen, Fred, 'The end of photography as we have known it', in *Photovideo: Photography in the age of the computer*, ed. Paul Wombell (London, 1991).
Robins, Kevin, 'Into the image: visual technologies and vision cultures', in *Photovideo: Photography in the age of the computer*, ed. Paul Wombell (London, 1991).
Rostow, Walt, *The Stages of Growth: A Non-Communist Manifesto*, 1960 (Cambridge, 1990).
Rotter, Andrew, *The Path to Vietnam* (London, 1987).
Rowlandson, Mary, *The Captive*, 1682 (Tucson, Ariz., 1988).
Rumaihi, Muhammad, *Beyond Oil: Unity and Development in the Gulf*, 1983, trans. James Dickens (London, 1986).
Rutherford, Jonathan, 'Who's That Man?', in *Male Order: Unwrapping Masculinity*, eds Rowena Chapman and Jonathan Rutherford (London, 1988).
Said, Edward, *Orientalism: Western Conceptions of the Orient*, 1978 (Harmondsworth, 1985).
—— *The Question of Palestine* (New York, 1980).
—— *Covering Islam: How the media and the experts determine how we see the rest of the world* (London, 1981).
—— *Culture and Imperialism* (London, 1994).
Sale, Kirkpatrick, *Power Shift: The Rise of the Southern Rim and Its Challenge to the Eastern Establishment* (New York, 1975).
—— *The Conquest of Paradise: Christopher Columbus and the Columbian Legacy* (London, 1991).
Samuel, Raphael and Thompson, Paul, eds, *The Myths We Live By* (London, 1990).
Sapir, Edward, 'Language', 1949, in *Sociological Perspectives*, eds Kenneth Thompson and Jeremy Tunstall (Harmondsworth, 1971).
—— *Culture, Language and Personality: Selected Essays* (Berkeley, Calif., 1962).
Saxton, Alexander, *The Rise and Fall of the White Republic: Class Politics and Mass Culture in Nineteenth-Century America* (London, 1990).
Schulz, Juergen, 'Jacopo de Barbari's View of Venice: Map Making, City Views, and Moralized Geography Before the Year 1500', *The Art Bulletin*, vol. 60, 1978.
Schwartz, Seymour and Ehrenberg, Ralph, *The Mapping of America* (New York, 1980).
Segal, Charles and Stineback, David, *Puritans, Indians, and Manifest Destiny* (New York, 1977).
Shawcross, William, *Sideshow: Kissinger, Nixon and the Destruction of Cambodia* (London, 1986).

Sheehan, Neil, *A Bright Shining Lie: John Paul Vann and America in Vietnam* (London, 1990).
Shirley, Rodney, *The Mapping of the World: Early Printed World Maps 1472–1700* (London, 1985).
Sinnhuber, K.A., 'The Representation of Disputed Political Boundaries in General Atlases', *Cartographic Journal*, vol. 1, no. 2, December 1964.
Slotkin, Richard, *Regeneration Through Violence: The Mythology of the American Frontier, 1600–1860* (Middletown, Conn., 1973).
—— *The Fatal Environment: The Myth of the Frontier in the Age of Industrialization 1800–1890* (New York, 1985).
Smith, David, *Maps and Plans for the Local Historian and Collector* (London, 1988).
Smith, Henry Nash, *Virgin Land: The American West as Symbol and Myth* (New York, 1950).
Smith, Michael, 'Selling the Moon', in *The Culture of Consumption: Critical Essays in American History 1880–1980*, eds R.W. Fox and T.J. Jackson Lears (New York, 1983).
Soja, Edward, *Postmodern Geographies: The Reassertion of Space in Critical Social Theory* (London, 1989).
Speed, John, *A Prospect of the Most Famous Parts of the World* (London, 1627).
Speier, Hans, 'Magic Geography', *Social Research*, vol. 8, no. 2, May 1941.
Stanislawski, Dan, 'The origin and spread of the grid-pattern town', *Geographical Review*, vol. 36, January 1946.
Stearn, William, 'Notes on Linnaeus's "Genera Plantarum"', introduction to Carl Linnaeus, *Genera Plantarum*, 1754 (Codicote, Herts., 1960).
Stein, Gertrude, *Picasso* (London, 1938).
Stilgoe, John, *Common Landscape of America, 1580 to 1845* (New Haven, Conn., 1982).
Swift, Graham, *Waterland* (London, 1984).
Swift, Jonathan, 'On Poetry: A Rapsody', in *The Poems of Jonathan Swift*, vol. II, ed. Harold Williams (Oxford, 1958).
Theweleit, Klaus, *Male Fantasies I: Women, floods, bodies, history*, 1977, trans. Stephen Conway (Cambridge, 1987).
Thrower, Norman, *Maps and Man: An Examination of Cartography in Relation to Culture and Civilization* (Englewood Cliffs, NJ, 1972).
Todorov, Tsvetan, *The Conquest of America: The Question of the Other*, 1982, trans. Richard Howard (New York, 1984).
Tooley, R.V., *Maps and Map-Makers* (London, 1949).
Tuan, Yi-Fu, *Topophilia: A Study of Environmental Perception, Attitudes, and Values* (New York, 1974).
Turnbull, Colin, *The Forest People*, 1961 (London, 1984).
Turner, Frederick J., 'The Significance of the Frontier in American History', 1893, in *Frontier and Section, Selected Essays of Frederick Jackson Turner* (Englewood Cliffs, NJ, 1961).
—— 'The Problem of the West', 1896, in *Frontier and Section.*
Turner, Frederick, *Beyond Geography* (New York, 1980).
Van Gennep, Arnold, *The Rites of Passage*, 1980, trans. Monika Vizedom and Gabrielle Caffee (London, 1960).
Veblen, Thorstein, *The Theory of the Leisure Class* (London, 1925).

Vico, Giambattista, *The New Science of Giambattista Vico*, 1744, trans. Thomas Bergin and Max Fisch (Ithaca, NY, 1968).
Warner, Rex, *The Aerodrome*, 1941 (Oxford, 1982).
Webb, James, *Space Age Management* (New York, 1969).
Weems, Mason, *The Life of Washington*, 1809, ed. Marcus Cunliffe (Cambridge, Mass., 1962).
West, Nathanael, *The Day of the Locust*, 1939, in *The Complete Works of Nathanael West* (London, 1983).
White, Hayden, 'The Historical Text as Literary Artefact', in *The Writing of History: Literary Form and Historical Understanding*, eds Robert Canary and Henry Kozicki (Madison, Wis., 1978).
White, Leslie, *The Pueblo of Santa Ana, New Mexico*, American Anthroplogical Association, Memoir 60, vol. 44, no. 4, 1942.
White, Richard, *Inventing Australia: Images and Identity, 1688–1980* (Sydney, 1981).
Whorf, Benjamin, *Language, Thought and Reality: Selected Writings of Benjamin Lee Whorf*, ed. John Carroll (Cambridge, Mass., 1956).
—— 'An American Indian Model of the Universe', in *Language, Thought and Reality*.
Williamson, Ray, *Living in the Sky: The Cosmos of the American Indian* (Norman, Okla., 1984).
Wilson, Alexander, *The Culture of Nature* (Oxford, 1992).
Wilson, Andrew, *The Bomb and the Computer* (London, 1968).
Winterson, Jeanette, *Sexing the Cherry* (London, 1989).
Wolfe, Tom, *The Right Stuff* (London, 1981).
Wolff, Janet, *The Social Production of Art* (London, 1981).
Wombell, Paul, ed., *Photovideo: Photography in the age of the computer* (London, 1991).
Wright, Stephen, *Meditations in Green*, 1983 (London, 1985).
Yergin, Daniel, *Shattered Peace: The Origins of the Cold War and the National Security State* (London, 1978).

Index